Professional English for Artificial Intelligence

# 人工智能专业英语

吕云翔 ◎ 编著

清华大学出版社
北京

## 内容简介

本书是面向人工智能专业英语课程的教材，全书共分为10个单元，分别为人工智能介绍、知识表示和推理、不确定性推理、人工智能的搜索方法、机器学习、人工神经网络、深度学习、强化学习、计算机视觉和自然语言处理。本书信息量大，知识性强，注重英语的听、说、读、写、译能力的全面培养和实际应用。各单元内容均分为阅读与翻译、写作、听与说三大部分；采用场景式教学和体验式学习相结合的方式，融合了角色扮演、多人会话和小组讨论等行之有效的训练方法。

本书适合国内各类高校智能科学与技术、计算机科学与技术、软件工程等相关专业教学之用，也可作为其相关专业或IT领域人员的自学参考用书。

本书封面贴有清华大学出版社防伪标签，无标签者不得销售。
版权所有，侵权必究。举报：010-62782989，beiqinquan@tup.tsinghua.edu.cn。

**图书在版编目(CIP)数据**

人工智能专业英语/吕云翔编著．—北京：清华大学出版社，2021.5（2024.6重印）
（清华科技大讲堂丛书）
ISBN 978-7-302-56900-8

Ⅰ．①人… Ⅱ．①吕… Ⅲ．①人工智能-英语-高等学校-教材 Ⅳ．①TP18

中国版本图书馆 CIP 数据核字(2020)第 226843 号

策划编辑：魏江江
责任编辑：王冰飞　李　晔
封面设计：刘　键
责任校对：徐俊伟
责任印制：沈　露

出版发行：清华大学出版社
　　　　　网　　址：https://www.tup.com.cn,https://www.wqxuetang.com
　　　　　地　　址：北京清华大学学研大厦 A 座　　邮　编：100084
　　　　　社 总 机：010-83470000　　　　　　　　邮　购：010-62786544
　　　　　投稿与读者服务：010-62776969，c-service@tup.tsinghua.edu.cn
　　　　　质量反馈：010-62772015，zhiliang@tup.tsinghua.edu.cn
　　　　　课件下载：https://www.tup.com.cn,010-83470236
印 装 者：三河市龙大印装有限公司
经　　销：全国新华书店
开　　本：185mm×260mm　　　印　张：14.75　　　字　数：369 千字
版　　次：2021 年 5 月第 1 版　　　　　　　　　印　次：2024 年 6 月第 4 次印刷
印　　数：3301～4100
定　　价：49.80 元

产品编号：090115-01

# Preface 前言

英语是全球 IT 行业的行业语言,英语技能是 IT 行业最基本的技能之一,因此熟练掌握相关英语技能对于发展职业生涯具有积极的影响。

本书是按照最新的《大学英语教学大纲》对专业英语的要求,为开设人工智能专业英语课程而编写的教材。在满足人工智能专业英语教学的同时,注重实际应用与调动学习兴趣。本书选材广泛,内容丰富。全书共分为 10 个单元,分别为人工智能介绍、知识表示和推理、不确定性推理、人工智能的搜索方法、机器学习、人工神经网络、深度学习、强化学习、计算机视觉和自然语言处理。

本书在对话场景的编排上以 3 位计算机专业的大学本科生 Mark、Henry 和 Sophie 的学习生活为主要背景,他们交流的话题围绕各章主题展开,并在对话中丰富各章主题,将全书内容巧妙地联系在一起。

本书信息容量大,知识性强,注重英语的听、说、读、写、译能力的全面培养和实际应用。各单元内容均分为阅读与翻译、写作、听说三大部分。

本书采用场景式教学和体验式学习相结合的方式,教材中设计的听力、口语融合了角色扮演、多人会话和小组讨论等行之有效的训练方法,能较好地满足课堂教学的需要。

另外,本书配有丰富的教辅资源,包括听与说录音、参考译文、练习答案,并提供授课 PPT。

本书建议教学时长为 48 学时或 32 学时(可根据具体情况进行适当的裁剪)。理论授课包括课堂讲解、讨论、练习等必要的课内教学环节。建议授课时间比例为:阅读部分 40%,写作部分 20%,听与说部分 40%。

---

**资源下载提示**

**课件等资源**:扫描封底的"课件下载"二维码,在公众号"书圈"下载。

**扩展资源**:扫描目录上方的二维码下载。

**在线听力**:扫描封底刮刮卡中的二维码,再扫描书中相应章节中的二维码可以在线学习。

---

本书由吕云翔编著,曾洪立参与了部分内容的编写并进行了素材整理及配套资源制作等。

在本书的编写过程中，得到了美国 Auburn 大学的 Yvonne Williams 女士的指导，在此表示衷心的感谢。感谢高峻逸和杜予同为本书提供的大力帮助。

本书试图融合听、说、读、写、译各项技能的训练，书中难免会有不尽如人意之处，敬请专家与读者不吝赐教，以使该书臻于完善。

编　者

2021 年 5 月于北京

# Contents 目录

扩展资源

**Unit 1  Introduction to Artificial Intelligence** ················ 001

Part 1  Reading & Translating ················ 002
       Section A: The Turing Test ················ 002
       Section B: Benefits and Risks of Artificial Intelligence ················ 006
Part 2  Simulated Writing: Uncovering the Secrets of Clear Writing (I) ············ 012
Part 3  Listening & Speaking ················ 015
       Dialogue: Artificial Intelligence 🎧 ················ 015
       Listening Comprehension: Thinking Machines 🎧 ················ 019
       Dictation: Intelligent Agent 🎧 ················ 019

**Unit 2  Knowledge Representation and Reasoning** ················ 021

Part 1  Reading & Translating ················ 022
       Section A: Representing and Manipulating Knowledge ················ 022
       Section B: Learning ················ 026
Part 2  Simulated Writing: Uncovering the Secrets of Clear Writing (II) ············ 030
Part 3  Listening & Speaking ················ 034
       Dialogue: Knowledge Representation and Reasoning 🎧 ················ 034
       Listening Comprehension: Logical Reasoning 🎧 ················ 036
       Dictation: Semantic Networks 🎧 ················ 037

**Unit 3  Reasoning with Uncertainty** ················ 039

Part 1  Reading & Translating ················ 040
       Section A: Reasoning with Uncertainty ················ 040
       Section B: Probabilistic Reasoning ················ 045
Part 2  Simulated Writing: Communicating with Social Media ················ 051
Part 3  Listening & Speaking ················ 055
       Dialogue: Reasoning with Uncertainty 🎧 ················ 055

　　　　Listening Comprehension：Fuzzy Logic 🎧 ·················· 056
　　　　Dictation：Bayesian Network 🎧 ······················· 057

| Unit 4 | Search Methods in Artificial Intelligence ············ **059** |

Part 1　Reading & Translating ································· 060
　　　　Section A：Heuristic Search ···························· 060
　　　　Section B：Genetic Algorithms ·························· 064
Part 2　Simulated Writing：Using Presentation Software to Write ········ 068
Part 3　Listening & Speaking ··································· 074
　　　　Dialogue：Search Methods in Artificial Intelligence 🎧 ······ 074
　　　　Listening Comprehension：A* Search 🎧 ················· 076
　　　　Dictation：Heuristic Search Techniques 🎧 ··············· 077

| Unit 5 | Machine Learning ································ **079** |

Part 1　Reading & Translating ································· 080
　　　　Section A：Decision Tree in Machine Learning ············· 080
　　　　Section B：K-means Clustering Algorithm and Example ······ 084
Part 2　Simulated Writing：Developing Reports and Proposals(I) ······ 090
Part 3　Listening & Speaking ··································· 093
　　　　Dialogue：Machine Learning 🎧 ························ 093
　　　　Listening Comprehension：Supervised Learning 🎧 ········· 097
　　　　Dictation：Unsupervised Learning 🎧 ···················· 098

| Unit 6 | Artificial Neural Networks ························ **099** |

Part 1　Reading & Translating ································· 100
　　　　Section A：Artificial Neural Networks ····················· 100
　　　　Section B：Handwritten Digit Recognition ················· 104
Part 2　Simulated Writing：Developing Reports and Proposals(II) ····· 109
Part 3　Listening & Speaking ··································· 112
　　　　Dialogue：Artificial Neural Network 🎧 ·················· 112
　　　　Listening Comprehension：Training Artificial Neural Networks 🎧 ···· 114
　　　　Dictation：Applications of Neural Networks 🎧 ············ 115

| Unit 7 | Deep Learning ································· **117** |

Part 1　Reading & Translating ································· 118
　　　　Section A：Deep Learning, Machine Learning, and AI ········ 118
　　　　Section B：Convolutional Neural Network ················· 123

# Contents

| | | |
|---|---|---|
| Part 2 | Simulated Writing: Writing Professional Letters (I) | 127 |
| Part 3 | Listening & Speaking | 132 |
| | Dialogue: Deep Learning 🎧 | 132 |
| | Listening Comprehension: Generative Adversarial Network 🎧 | 135 |
| | Dictation: Recurrent Neural Network 🎧 | 136 |

## Unit 8  Reinforcement Learning — 138

| | | |
|---|---|---|
| Part 1 | Reading & Translating | 139 |
| | Section A: Reinforcement Learning, Deep Learning's Partner | 139 |
| | Section B: AlphaGo Zero: Starting from Scratch | 143 |
| Part 2 | Simulated Writing: Writing Professional Letters (II) | 147 |
| Part 3 | Listening & Speaking | 151 |
| | Dialogue: Reinforcement Learning 🎧 | 151 |
| | Listening Comprehension: Deep Reinforcement Learning 🎧 | 154 |
| | Dictation: Reinforcement Learning Challenges 🎧 | 155 |

## Unit 9  Computer Vision — 156

| | | |
|---|---|---|
| Part 1 | Reading & Translating | 157 |
| | Section A: What's the Difference between Computer Vision, Image Processing and Machine Learning? | 157 |
| | Section B: Using Vision for Controlling Movement | 162 |
| Part 2 | Simulated Writing: Writing for Employment (I) | 166 |
| Part 3 | Listening & Speaking | 169 |
| | Dialogue: Computer Vision 🎧 | 169 |
| | Listening Comprehension: Pattern Recognition 🎧 | 172 |
| | Dictation: Artificial Intelligence, Machine Learning, Deep Learning, and Computer Vision…What is the Difference? 🎧 | 173 |

## Unit 10  Natural Language Processing — 175

| | | |
|---|---|---|
| Part 1 | Reading & Translating | 176 |
| | Section A: Language Understanding | 176 |
| | Section B: Natural Language Processing vs. Machine Learning vs. Deep Learning | 180 |
| Part 2 | Simulated Writing: Writing for Employment (II) | 184 |
| Part 3 | Listening & Speaking | 189 |
| | Dialogue: Natural Language Processing 🎧 | 189 |
| | Listening Comprehension: Speech Recognition 🎧 | 192 |

Dictation：Natural Language Understanding ........................................ 193

**Glossary** .................................................... **195**

**Abbreviations** ............................................. **209**

**Answers** .................................................... **210**

**Bibliography** .............................................. **224**

Complementary Materials
人工智能补充阅读材料

# Unit 1
## Introduction to Artificial Intelligence

# Part 1

## Reading & Translating

### Section A: The Turing Test

In 1950, English mathematician Alan Turing wrote a **landmark** paper that asked the question: Can machines think? After carefully defining terms such as "intelligence" and "thinking," he ultimately concluded that we would eventually be able to create a computer that thinks. But then he asked another question: How will we know when we've succeeded?

His answer to that question came to be called the Turing test (Figure 1-1), which is used to **empirically** determine whether a computer has achieved intelligence. The test is based on whether a computer could fool a human into believing that the computer is another human.

Figure 1-1  Turing test

In the past the Turing test has served as a benchmark in measuring progress in the field of artificial intelligence. Today the significance of the Turing test has **faded** although it remains an important part of the artificial intelligence **folklore**. Turing's proposal was to allow a human, whom we call the **interrogator**, to communicate with a test subject by means of a typewriter system without being told whether the test subject was a human or a machine. In this environment, a machine would be declared to behave intelligently if the interrogator was not able to distinguish it from a human. Turing predicted that by the year 2000 machines would have a 30 percent chance of passing a five-minute Turing test—a **conjecture** that **turned out to be** surprisingly accurate.

One reason that the Turing test is no longer considered to be a meaningful measure of intelligence is that an **eerie** appearance of intelligence can be produced with relative ease. A well-known example arose as a result of the program DOCTOR (a version of the more general system called ELIZA) developed by Joseph Weizenbaum in the mid-1960s. This interactive program was designed to **project** the **image** of a Rogerian analyst conducting a psychological interview; the computer played the role of the analyst while the user played the patient. Internally, all that DOCTOR did was to restructure the statements made by the patient according to some **well-defined** rules and direct them back to the patient. For example, in response to the statement "I am tired today," DOCTOR might have replied with "Why do you think you're tired today? " If DOCTOR was unable to recognize the sentence structure, it merely responded with something like "Go on" or "That's very interesting."

Weizenbaum's purpose in developing DOCTOR dealt with the study of natural language communication. The subject of psychotherapy merely provided an environment in which the program could "communicate." To Weizenbaum's **dismay**, however, several psychologists proposed using the program for actual psychotherapy. (The Rogerian thesis is that the patient, not the analyst, should lead the discussion during the **therapeutic session**, and thus, they **argued**, a computer could possibly conduct a discussion as well as a therapist could.) Moreover, DOCTOR projected the image of comprehension so strongly that many who "communicated" with it became **subservient** to the machine's question-and-answer dialogue. In a sense, DOCTOR passed the Turing test. The result was that ethical, as well as technical, issues were raised, and Weizenbaum became an advocate for maintaining human dignity in a world of advancing technology.

More recent examples of Turing test "successes" include Internet viruses that carry on "intelligent" dialogs with a human victim in order to **trick** the human into dropping his or her malware guard. Moreover, phenomena similar to Turing tests occur in the context of computer games such as chess-playing programs. Although these programs select moves merely by applying brute-force techniques, humans competing against the computer often experience the sensation that the machine possesses creativity and even a personality. Similar sensations occur in robotics where machines have been built with physical attributes that project intelligent characteristics. Examples include toy robot dogs that project **adorable** personalities merely by **tilting** their heads or lifting their ears in response to a sound.

Some AI researchers assert that true artificial intelligence will not exist until we have achieved strong equivalence—that is, until we create a machine that processes information as the human mind does.

New York **philanthropist** Hugh Loebner organized the first formal **instantiation** of the Turing test. This competition has been run annually since 1991. A grand prize of $100 000 and a solid gold medal will be awarded for the first computer whose responses are indistinguishable from a human's. So far the grand prize remains **up for grabs**. A prize of $2000 and a bronze medal is awarded each year for the computer that is determined to be the most human-like, relative to the rest of the competition that year. The Loebner prize contest has become an important annual event for computing enthusiasts interested in artificial intelligence.

Various programs, often referred to as Chatbots, have been developed to perform this kind of conversational interaction between a computer and a person. Many are available through the World Wide Web and focus on a particular topic. Depending on how well they are designed, these programs can carry on a reasonable conversation. In most cases, though, it doesn't take long for the user to discover **awkward** moments in the conversation that **betray** the fact that a human mind is not determining the responses.

## Words

| | |
|---|---|
| landmark[ˈlændmɑːk] *adj.* 有重大意义或影响的 | argue[ˈɑːgjuː] *v.* 证明, 说服 |
| empirically[imˈpirikli] *adv.* 以经验为主地 | subservient[səbˈsɜːviənt] *adj.* 有用的, 有帮助的 |
| fade[feid] *adj.* 平淡的, 乏味的 | trick[trik] *v.* 欺骗, 哄骗 |
| folklore[ˈfəuklɔː(r)] *n.* 民间传说 | adorable[əˈdɔːrəbl] *adj.* 可爱的, 值得敬重的 |
| interrogator[inˈterəgeitə(r)] *n.* 讯问者, 询问机 | tilt[tilt] *v.* 翘起, 倾斜 |
| conjecture[kənˈdʒektʃə(r)] *n.* 推测, 猜想 | philanthropist[fiˈlænθrəpist] *n.* 慈善家, 乐善好施的人 |
| eerie[ˈiəri] *adj.* 可怕的, 怪异的 | instantiation[instænʃiˈeiʃən] *n.* 实例化, 例示 |
| project[ˈprɔdʒekt] *v.* 表现, 设计 | |
| image[ˈimidʒ] *n.* 生动的描绘, 概念 | |
| well-defined 明确的 | awkward[ˈɔːkwəd] *adj.* 尴尬的, 不合适的 |
| dismay[disˈmei] *n.* 沮丧, 灰心 | |
| therapeutic[θerəˈpjuːtik] *adj.* 治疗的, 医疗的 | betray[biˈtrei] *v.* 背叛(原则或信仰) |

## Phrases

turn out to be　证明是, 结果是, 原来是
up for grabs　大家有份

# Unit 1  Introduction to Artificial Intelligence

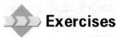 **Exercises**

**I. Read the following statements carefully, and decide whether they are true (T) or false (F) according to the text.**

____ 1. Alan Turing organized the first formal instantiation of the Turing test.

____ 2. In 1991, English mathematician Alan Turing wrote a landmark paper that asked the question: Can machines think?

____ 3. Program DOCTOR was developed by Alan Turing.

____ 4. After Alan Turing asked the question: Can machines think? then he asked another question: How will we know when we've succeeded?

____ 5. The Turing test is based on whether a computer could tease a human to make him believe that the computer is another human.

**II. Choose the best answer to each of the following questions according to the text.**

1. Who organized the first formal instantiation of the Turing test? (    )
   A. Alan Turing
   B. Hugh Loebner
   C. Weizenbaum
   D. All of the above

2. When did Alan Turing write a landmark paper that asked the question: Can machines think? (    )
   A. 1991
   B. 1960
   C. 1950
   D. None of the above

3. Which of the following is based on whether a computer could fool a human to make him believe that the computer is another human? (    )
   A. DOCTOR
   B. Chatbot
   C. ELIZA
   D. Turing test

**III. Fill in the numbered spaces with the words or phrases chosen from the box. Change the forms where necessary.**

| while | outperform | seem | research | however |
|---|---|---|---|---|
| know | instead | performance | approach | apply |

**Behavior-Based Intelligence**

Early work in artificial intelligence ____1____ the subject in the context of explicitly writing programs to simulate intelligence. ___2___, many argue today that human intelligence is not based on the execution of complex programs but ___3___ by simple stimulus-response functions that have evolved over generations. This theory of "intelligence" is ___4___ as behavior-based intelligence because "intelligent" stimulus response functions appear to be the result of behaviors that caused certain individuals to survive and reproduce ___5___ others did not.

Behavior-based intelligence ___6___ to answer several questions in the artificial intelligence community such as why machines based on the von Neumann architecture easily ___7___ humans in computational skills but struggle to exhibit common sense. Thus behavior-based intelligence promises to be a major influence in artificial intelligence ___8___. As we know, behavior-based techniques have been ___9___ in the field of artificial neural networks to teach neurons to behave in desired ways, in the field of genetic algorithms to provide an alternative to the more traditional programming process, and in robotics to improve the ___10___ of machines through reactive strategies.

IV. Translate the following passage into Chinese.

**Artificial Intelligence in the Palm of Your Hand**

Artificial intelligence techniques are increasingly showing up in smartphone applications. For example, Google has developed Google Goggles, a smartphone application providing a visual search engine. Just take a picture of a book, landmark, or sign using a smartphone's camera and Goggles will perform image processing, image analysis, and text recognition, and then initiate a Web search to identify the object. If you are an English speaker visiting in France, you can take a picture of a sign, menu, or other text and have it translated to English. Beyond Goggles, Google is actively working on voice-to-voice language translation. Soon you will be able to speak English into your phone and have your words spoken in Spanish, Chinese, or another language. Smartphones will undoubtedly get smarter as AI continues to be utilized in innovative ways.

## Section B: Benefits and Risks of Artificial Intelligence

From Siri to self-driving cars, Artificial Intelligence (AI) is progressing rapidly. While science fiction often portrays AI as robots with human-like characteristics, AI can **encompass** anything from Google's search algorithms to IBM's Watson to autonomous weapons.

Artificial intelligence today is properly known as narrow AI (or weak AI), in that it is designed to perform a narrow task (e.g. only facial recognition or only Internet searches or only driving a car). However, the long-term goal of many researchers is to create general AI (AGI or strong AI). While narrow AI may **outperform** humans at whatever its specific task is, like playing chess or solving equations, AGI would

## Unit 1  Introduction to Artificial Intelligence

outperform humans at nearly every cognitive task.

In the near term, the goal of keeping AI's impact on society benefits motivates research in many areas, from economics and law to technical topics such as verification, validity, security and control. Whereas it may be little more than a minor **nuisance** if your laptop crashes or gets hacked, it becomes all the more important that an AI system does what you want it to do if it controls your car, your airplane, your **pacemaker**, your automated trading system or your power grid.[1] Another short-term challenge is preventing a devastating arms race in **lethal** autonomous weapons.

In the long term, an important question is what will happen if the **quest** for strong AI succeeds and an AI system becomes better than humans at all cognitive tasks. As pointed out by I. J. Good in 1965, designing smarter AI systems is itself a cognitive task. Such a system could potentially undergo **recursive** self-improvement, triggering an intelligence explosion leaving human **intellect** far behind. By inventing revolutionary new technologies, such super-intelligence might help us **eradicate** war, disease, and poverty, and so the creation of strong AI might be the biggest event in human history. Some experts have expressed concern, though, that it might also be the last, unless we learn to **align** the goals of the AI **with** ours before it becomes super-intelligent.

There are some who question whether strong AI will ever be achieved, and others who insist that the creation of super-intelligent AI is guaranteed to be beneficial. In some artificial intelligent companies we recognize both of these possibilities, but also recognize the potential for an artificial intelligence system to intentionally or unintentionally cause great harm (Figure 1-2). We believe research today will help us better prepare for and prevent such potentially negative consequences in the future, thus enjoying the benefits of AI while avoiding pitfalls.

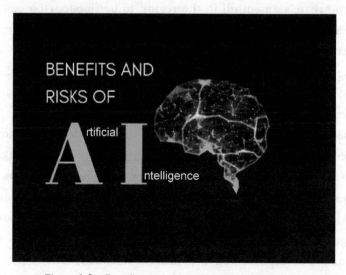

Figure 1-2  Benefits and risks of artificial intelligence

Most researchers agree that a super-intelligent AI is unlikely to exhibit human emotions like love or hate, and that there is no reason to expect AI to become intentionally **benevolent** or **malevolent**. Instead, when considering how AI might become a risk, experts think two scenarios most likely:

1. The AI is programmed to do something devastating: Autonomous weapons are artificial intelligence systems that are programmed to kill. In the hands of the wrong person, these weapons could easily cause mass **casualties**. Moreover, an AI **arms race** could **inadvertently** lead to an AI war that also results in mass casualties. To avoid being **thwarted** by the enemy, these weapons would be designed to be extremely difficult to simply "turn off," so humans could **plausibly** lose control of such a situation. This risk is one that's present even with narrow AI, but grows as levels of AI intelligence and autonomy increase.

2. The AI is programmed to do something beneficial, but it develops a destructive method for achieving its goal: This can happen whenever we fail to fully align the AI's goals with ours, which is **strikingly** difficult. If you ask an **obedient** intelligent car to take you to the airport as fast as possible, it might get you there chased by helicopters and **covered** in **vomit**, doing not what you wanted but **literally** what you asked for. If a super-intelligent system is **tasked** with an ambitious geo-engineering project, it might **wreak havoc** with our ecosystem as a **side effect**, and view human attempts to stop it as a threat to be met.

As these examples illustrate, the concern about advanced AI isn't malevolence but competence. A super-intelligent AI will be extremely good at accomplishing its goals, and if those goals aren't aligned with ours, we have a problem. You're probably not an evil ant-hater who **steps on** ants **out of malice**, but if you're in charge of a hydroelectric green energy project and there's an anthill in the region to be flooded, too bad for the ants. A key goal of AI safety research is to never place humanity in the position of those ants.

Stephen Hawking, Elon Musk, Steve Wozniak, Bill Gates, and many other big names in science and technology have recently expressed concern in the media and via open letters about the risks **posed** by AI, joined by many leading AI researchers. Why is the subject suddenly in the headlines?

The idea that the quest for strong AI would ultimately succeed **was long thought of as** science fiction, centuries or more away. However, thanks to recent breakthroughs, many AI milestones, which experts viewed as decades away merely five years ago, have now been reached, making many experts take seriously the possibility of super-intelligence in our lifetime. While some experts still guess that human-level AI is centuries away, some AI researchers guessed that it would happen before 2060. Since it may take decades to complete the required safety research, it is **prudent** to start it now.

Because AI has the potential to become more intelligent than any human, we have no **surefire** way of predicting how it will behave. We can't use past technological

# Unit 1  Introduction to Artificial Intelligence

developments as much of a basis because we've never created anything that has the ability to, **wittingly** or **unwittingly**, **outsmart** us. The best example of what we could face may be our own evolution. People now control the planet, not because we're the strongest, fastest or biggest, but because we're the smartest. If we're no longer the smartest, are we assured to remain in control?

## Words

encompass[ɪnˈkʌmpəs] v. 包含,包括,涉及(大量事物)
outperform[ˌaʊtpəˈfɔːm] v. (效益上)超过,胜过
nuisance[ˈnjuːsns] n. 损害,麻烦事
pacemaker[ˈpeɪsmeɪkə(r)] n. 心脏起搏器
lethal[ˈliːθl] adj. 致命的,致死的
quest[kwest] n. 追求,寻找
recursive[rɪˈkɜːsɪv] adj. 循环的
intellect[ˈɪntəlekt] n. 智力,才智
eradicate[ɪˈrædɪkeɪt] v. 根除,消灭,杜绝
benevolent[bəˈnevələnt] adj. 仁慈的,慈善的
malevolent[məˈlevələnt] adj. 恶毒的,有恶意的
casualty[ˈkæʒuəlti] n. (战争或事故的)伤员,遇难者,受害者
inadvertently[ˌɪnədˈvɜːtntli] adv. 无意地,不经意地

thwart[θwɔːt] v. 挫败,反对
plausibly[ˈplɔːzəbli] adv. 振振有词地,似乎合理地,似是而非地,似乎可信地
strikingly[ˈstraɪkɪŋli] adv. 显著地,突出地
obedient[əˈbiːdiənt] adj. 顺从的,服从的
cover[ˈkʌvə(r)] v. 行走(一段路程)
vomit[ˈvɒmɪt] n. 呕吐
literally[ˈlɪtərəli] adv. 真正地,确实地,简直
task[tɑːsk] v. 派给某人(任务)
pose[pəʊz] v. 造成,形成
prudent[ˈpruːdnt] adj. 谨慎的,慎重的
surefire[ˈʃʊəfaɪə] adj. 十拿九稳的,一定成功的
wittingly[ˈwɪtɪŋli] adv. 有意地
unwittingly[ʌnˈwɪtɪŋli] adv. 不经意地
outsmart[ˌaʊtˈsmɑːt] v. 比……更聪明,用计谋打败

## Phrases

align with    使一致,对准
arms race    军备竞赛
wreak havoc    造成严重破坏,肆虐
side effect    (药物的)副作用,意外的连带后果
step on    踩上……,踏上……
out of malice    出于恶意
be thought of as    把……看作,被认为是

009

**Notes**

[1] **Original**:Whereas it may be little more than a minor nuisance if your laptop crashes or gets hacked,it becomes all the more important that an AI system does what you want it to do if it controls your car,your airplane,your pacemaker,your automated trading system or your power grid.

**Translation**:如果你的笔记本电脑崩溃或被黑客入侵,那可能只是些小麻烦,但如果 AI 系统控制着你的汽车、飞机、心脏起搏器、自动交易系统或电网,那么它们就会变成大问题。

 **Exercises**

I. Read the following statements carefully,and decide whether they are true(T)or false(F)according to the text.

　　____ 1. The concern about advanced AI isn't malevolence but benevolent.

　　____ 2. Human-level AI would happen before 2030.

　　____ 3. Artificial intelligence today is properly known as strong AI.

　　____ 4. In 1965 Bill Gates point out that designing smarter AI systems is itself a cognitive task.

　　____ 5. AI(or weak AI)is designed to perform a narrow task like facial recognition.

II. Choose the best answer to each of the following questions according to the text.

1. Which of the following description is not right?(　　)

　　A. Strong AI would outperform humans at nearly every cognitive task.

　　B. Because AI has the potential to become more intelligent than any human,we have no surefire way of predicting how it will behave.

　　C. While some experts still guess that human-level AI is centuries away,some AI researchers guessed that it would happen before 2030.

　　D. In the long term,an important question is what will happen if the quest for strong AI succeeds and an AI system becomes better than humans at all cognitive tasks.

2. Which of the following is a task performed by weak AI?(　　)

　　A. Internet searches

　　B. Facial recognition

　　C. Driving a car

　　D. All of the above

3. Which of the following description is right?(　　)

　　A. AI today can exhibit human emotions like love or hate.

# Unit 1  Introduction to Artificial Intelligence

B. AI today can exhibit human moral characters like benevolent or malevolent.
C. Strong AI might help us eradicate war, disease, and poverty.
D. All of the above

III. Fill in the numbered spaces with the words or phrases chosen from the box. Change the forms where necessary.

> inherent  individual  however  observe  proponent
> exhibit  characteristics  program  debate  resolve

**Strong AI versus Weak AI**

The conjecture that machines can be __1__ to exhibit intelligent behavior is known as weak AI and is accepted, to varying degrees, by a wide audience today. __2__ the conjecture that machines can be programmed to possess intelligence and, in fact, consciousness, which is known as strong AI, is widely __3__. Opponents of strong AI argue that a machine is __4__ different from a human and thus can never feel love, tell right from wrong, and think about itself in the same way that a human does. However, __5__ of strong AI argue that the human mind is constructed from small components that __6__ are not human and are not conscious but, when combined, are. Why, they argue, would the same phenomenon not be possible with machines?

The problem in __7__ the strong AI debate is that such attributes as intelligence and consciousness are internal __8__ that cannot be identified directly. As Alan Turing pointed out, we credit other humans with intelligence because they behave intelligently—even though we cannot __9__ their internal mental states. Are we, then, prepared to grant the same latitude to a machine if it __10__ the external characteristics of consciousness?

IV. Translate the following passage into Chinese.

**Physical Agents**

A physical agent (robot) is a programmable system that can be used to perform a variety of tasks. Simple robots can be used in manufacturing to do routine jobs such as assembling, welding, or painting. Some organizations use mobile robots that do delivery jobs such as distributing mail or correspondence to different rooms. There are mobile robots that are used underwater for prospecting for oil.

A humanoid robot is an autonomous mobile robot that is supposed to behave like a human. Although humanoid robots are prevalent in science fiction, there is still a lot of work to do before such robots will be able to interact properly with their surroundings and learn from events that occur there.

# Part 2

# Simulated Writing：Uncovering the Secrets of Clear Writing（I）

好的作者不是天生的。他们通过不断地练习和对细节的关注来培养自己的技能。我们也可以通过同样的方式来成为一个好的作者——不断练习并关注细节。我们培养出来的能够清晰明了地写作的技能会使我们在整个职业生涯中受益。一些人误以为清晰地写作就是文档中没有一点儿错误（比如拼写或语法错误）。清晰地写作实际上是指能够在最好地满足读者的需求和兴趣的情况下进行沟通。一个书写良好的文档具有特定的目的、明晰的观点，将支持性和相关性信息组织得富有逻辑，并且语法正确。

**1. 使书面交流变得清晰明了**

各个领域的专业人士必须能够在写作时清晰、简要并完整地表达想法。如果书面表达不清晰或者缺少重要的细节，读者会迷惑不解并且不能恰当地给出回复。要想清晰地写作，确保完成以下任务：准备、撰写以及修订，如图1-3所示。

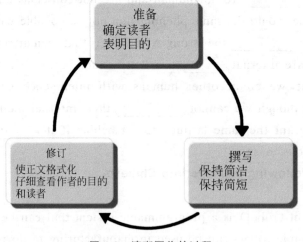

图1-3　清晰写作的过程

在开始写作之前，复习一下以下指导准则：

1) 了解读者

以识别典型的读者来开始写作工作。通常情况下，作者需要帮助读者明白文档的主题和想法，说服他们支持文档的观点，或者动员他们采取行动。如果刚开始就从读者的角度而不是作者自己的角度来思考某个主题，那么作者就可以成功。如果文档很长、很复杂，或者非常重要，可以起草一个读者概况，写作的时候可作为参考。

2) 与读者的经历和理解相关

理想情况下，作者的文档不会令读者感到惊讶。相反，作者应该增加读者的知识储备，并且和他们一起探索另一种想法。一个技能高超的作者可以确定他的读者所了解的主题并基于此来引入新的概念。考虑读者的经历可以帮助作者使写作更清晰明了，更易于理解，并且更具有相关性。

3) 明确目标

在开始写作之前，要确定文档的确切目的。为什么要写作并且期望获得什么？大多数专业写作的目的都是为了告知某事，例如宣布开会、概述决策，或者列出流程。许多企业文档的目的都是为了说服，例如为了说服经理或者同事，激励员工，或者刺激消费者采取行动。写作的时候，重复检查以确保每一个句子以及每一段话都有助于实现目标。

4) 保持简洁

专业的写作应该是高效的，也就是说，易于阅读和理解。使文字、句子和段落简短并切中要点。去掉模糊用语和不必要的文字可以使文档"瘦身"。密切注意长句子并认真复查任何长于两行的内容。表1-1列出了使作品清晰明了的注意事项。

表 1-1　使作品清晰明了的注意事项

| 指导准则 | 适合提到 | 尽量避免 |
| --- | --- | --- |
| 了解读者 | • 确定是写给同事、决策者还是消费者<br>• 考虑有多少读者已经知道文档的主题了<br>• 指出读者可以从这个文档中获得的好处 | • 忽略次要读者<br>• 因为文档短而跳过此步骤 |
| 与读者关联 | • 描述典型的读者<br>• 期待读者的反应<br>• 使文档符合读者<br>• 使用合适的词汇和语气<br>• 直接使用"你"来指代读者 | • 使用不适合读者的语言，比如术语或缩写<br>• 选择可能被理解为偏见的词汇<br>• 忘了语气<br>• 仅仅关注需求和目标 |
| 明确目标 | • 确定文档的目的<br>• 指出想要利用这个文档获得什么 | • 包含不能满足目标的句子 |
| 保持简洁 | • 使用简短、常见的词汇<br>• 使用短句子和段落 | • 使用模糊或者不必要的词语<br>• 包含长于两行的句子 |
| 使文档具有吸引力 | • 良好地组织并格式化文档，使其吸引读者 | • 将文档格式化为一个大段的文本 |

5) 使文档具有吸引力

读者会积极地确定是否去阅读作者已经准备的文档。影响该决定的一个重要因素就是作品看起来怎么样。那些使用清晰的书面语并使用了具有吸引力的布局的文档更吸引人，并且更有可能被人阅读。相反，大段的文档会吓跑读者，并且减少他们认真思考作品的机会。

总而言之，作品可以遵循以下指导准则：从读者角度出发，专注于能够满足目标的信息，并且仔细地选择语言。

**2. 掌握标点符号**

当和某人谈话时，说话的方式会传达很多意思。停顿、语调抑扬变化，以及语速等会帮助表达想法并使意思清晰。书面交流使用一组称作标点的符号来实现这些任务，并且帮助读者解释文字。最常用的标点符号是句号、逗号和问号。表1-2总结了使用逗号和冒号时的注意事项。图1-4展示了某些类型的标点实例。

表 1-2 使用逗号和冒号时的注意事项

| 标点符号元素 | 适合提到 | 尽量避免 |
|---|---|---|
| 逗号 | • 引出介绍性文字<br>例如，Generally, employees arrive on time<br>• 列出一系列条目<br>例如，She writes, copies, and prints the articles<br>• 使独立分句分开<br>例如，The Web page is colorful, but it does't provide much information | • 省略最后一个逗号，如果省略了，则会导致歧义<br>• 使用逗号将没有 FANBOYS 的独立分句分开 |
| 冒号 | • 引出一个列表、例子或者引用<br>例如，You'll receive the following items at the conference: handouts, samples, and exhibition passes<br>• 在商务信函中表示问候语的结束<br>例如，Dear Mr. Wolff:<br>• 将标签(号)或短标题和文字信息分开<br>例如，Subject: Budget decisions | • 插入一个冒号，除非一边的文字形成了一个完整的句子 |

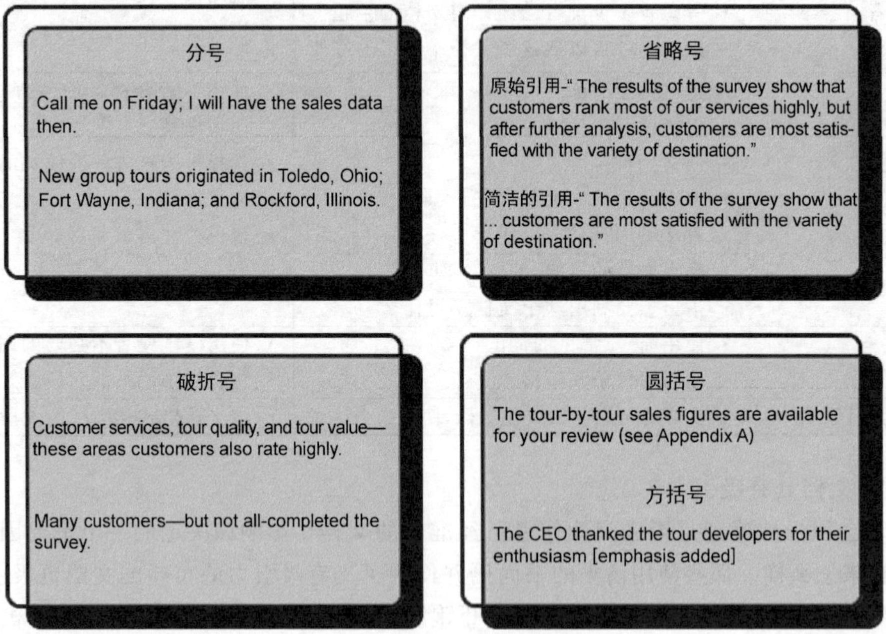

**分号**

Call me on Friday; I will have the sales data then.

New group tours originated in Toledo, Ohio; Fort Wayne, Indiana; and Rockford, Illinois.

**省略号**

原始引用-" The results of the survey show that customers rank most of our services highly, but after further analysis, customers are most satisfied with the variety of destination."

简洁的引用-" The results of the survey show that ... customers are most satisfied with the variety of destination."

**破折号**

Customer services, tour quality, and tour value—these areas customers also rate highly.

Many customers—but not all-completed the survey.

**圆括号**

The tour-by-tour sales figures are available for your review (see Appendix A)

**方括号**

The CEO thanked the tour developers for their enthusiasm [emphasis added]

图 1-4 某些类型的标点实例

1) 句号(。)和问号(?)

使用句号来结束一个完整的句子，甚至在项目符号列表或者编号列表中使用。使用问号来结束一个问句。

2) 逗号(,)

插入逗号是向读者展示哪些词语在句子中是同属于一类的。逗号的典型用法包括引出介绍性文字，列出 3 个或以上的一系列项目，使得用"for""and""not""but""or""yet"和"so"(这些都是连词，有时缩写为 FANBOYS)连接的独立分句分开。

3）分号（；）

当想要连接两个独立分句并且不使用像"and"或者"so"这样的连词，但要展示它们在意思上相关的时候使用分号。独立分句具有一个主语和一个动词，并且可以独立成句。

4）冒号（：）

使用冒号可以使读者意识到后面的信息是对当前想法的解释或提高。例如，一个冒号通常可以引出一个列表、例子或者引用。

5）省略号（…）

省略号表明已经省略了引用的一个或者多个单词。如果省略号出现在引用的句子的末尾，应在省略号之后加一个句号。

6）破折号（—）

插入破折号来引出或者强调句子的某一部分。一个单独的破折号强调紧随其后的内容；一对破折号强调包含在它们之间的内容。

7）圆括号（ ）和方括号 [ ]

使用圆括号插入某一关注，对句子的意思来说不是必需的。也可以在圆括号中包含附加的解释或者引用。当想要在段落中间插入自己的评论或观察结果时，使用方括号。这些评论通常用来解释或阐明正文中提到的想法。

转 30 页

# Part 3

## Listening & Speaking

### Dialogue：Artificial Intelligence

(*After class, Sophie & Henry are standing by the door, waiting for Mark.*)

**Henry**： Excuse me, Sophie. As you know Artificial Intelligence today is very hot. May I ask you some questions about AI?

**Sophie**： Sure. What can I do for you?[1]

**Henry**： What do you think of Artificial Intelligence?

**Sophie**： Let me see. To my understanding, the term Artificial Intelligence（AI）was **coined** in 1956, but AI has become more popular today thanks to increased data volumes, advanced algorithms, and improvements in computing power and storage.

[1] Replace with：
1. Can I help you?
2. May I help you?

在线音频

**Henry:** Well, could you please [2] talk about AI's history briefly?

**Sophie:** Of course. Early AI research in the 1950s explored topics like problem solving and symbolic methods. In the 1960s, the US Department of Defense took interest in this type of work and began training computers to **mimic** basic human reasoning. For example, the Defense Advanced Research Projects Agency (DARPA) completed street mapping projects in the 1970s. And DARPA produced intelligent personal assistants in 2003, long before Siri①, Alexa② or Cortana③ were **household** names.

**Henry:** So this means that those work impacted AI today?

**Sophie:** Absolutely. This early work paved the way for the automation and formal reasoning that we see in computers today, including decision support systems and smart search systems that can be designed to **complement** and **augment** human abilities.

[2] Replace with:
1. would you please
2. could you kindly

(When they are talking, Mark comes toward them.)

**Sophie & Henry:** Hi, Mark.

**Mark:** Hi, Henry and Sophie.

**Sophie:** You are just on time. Just before Henry was asking me about AI. I heard that you are quite familiar with AI.

**Mark:** A little bit. To my knowledge, while Hollywood movies and science fiction novels depict AI as human-like robots that take over the world, the current evolution of AI technologies isn't that scary—or quite that smart. Instead[3], AI has evolved to provide many specific benefits in every industry.

**Henry:** You mean specific benefits in every industry?

**Mark:** Yes. AI automates repetitive learning and discovery through data. But AI is different from hardware-driven, robotic automation. Instead of automating manual tasks, AI performs frequent, high-volume, computerized tasks reliably and without **fatigue**. For this type of automation, human inquiry is still essential to set up the system and ask the right questions.

[3] Replace with:
1. On the contrary
2. Rather

## Unit 1　Introduction to Artificial Intelligence

Sophie: I think AI adds intelligence to existing products. In most cases, AI will not be sold as an individual application. Rather, products you already use will be improved with AI capabilities, much like Siri was added as a feature to a new generation of Apple products. Automation, conversational platforms, **bots** and smart machines can be combined with large amounts of data to improve many technologies at home and in the workplace, from security intelligence to investment analysis.

Mark: You are right. AI adapts through progressive learning algorithms to let the data do the programming. AI finds structure and **regularities** in data so that the algorithm acquires a skill: The algorithm becomes a classifier or a predictor. So, just as the algorithm can teach itself how to play chess, it can teach itself what product to recommend next online. And the models adapt when given new data. Back propagation is an AI technique that allows the model to adjust, through training and added data, when the first answer is not quite right.

Henry: Ok, so what else?

Sophie: Well, AI analyzes more and deeper data using neural networks that have many hidden layers. Building a fraud detection system with five hidden layers was almost impossible a few years ago. All that has changed with incredible computer power and big data. You need lots of data to train deep learning models because they learn directly from the data. The more data you can feed them, the more accurate they become.

Mark: Besides, AI gets the most out of data. When algorithms are self-learning, the data itself can become intellectual property. The answers are in the data; you just have to apply AI to get them out. Since the role of the data is now more important than ever before, it can create a competitive advantage. If you have the best data in a competitive industry, even if everyone is applying similar techniques, the best data will win.

**Sophie**: Indeed, AI achieves incredible accuracy through deep neural networks—which was previously impossible. For example, your interactions with Alexa, Google Search and Google Photos are all based on deep learning—and they keep getting more accurate the more we use them. In the medical field, AI techniques from deep learning, image classification and object recognition can now be used to find cancer on **MRIs** with the same accuracy as highly trained **radiologists**.

**Henry**: Ok, I've got it. Sophie and Mark, thanks for your valuable knowledge.

**Sophie & Mark**: My pleasure.

## Exercises

Work in a group, and make up a similar conversation by replacing the statements with other expressions on the right side.

## Words

| | |
|---|---|
| coin[kɔin] v. 创造(新词,短语),杜撰 | 惊慌的 |
| mimic[ˈmimik] v. 模仿 | fatigue[fəˈtiːg] n. 疲劳,疲乏 |
| household[ˈhaushəuld] adj. 家喻户晓的 | bot[bɒt] n. 网上机器人,自动程序 |
| complement[ˈkɔmplim(ə)nt] v. 补足, 补助 | regularity[ˌregjuˈlærəti] n. 正则性,规律性,规则性 |
| augment[ɔːgˈment] v. 增加,增大 | radiologist[ˌreidiˈɔlədʒist] n. 放射科医生,放射线研究者 |
| scary[ˈskeəri] adj. (事物)可怕的,引起 | |

## Abbreviations

MRI  Magnetic Resonance Imaging  核磁共振成像

## Notes

① Siri 是 Speech Interpretation & Recognition Interface 的首字母缩写,原义为语音识别接口,是苹果公司在 iPhone、iPad、iPod Touch、HomePod 等产品上应用的一个语音助手,使用 Siri 用户可以通过手机读短信、介绍餐厅、询问天气、语音设置闹钟等。

② Alexa 是一家专门发布网站世界排名的网站。Alexa 每天在网上搜集超过 1000GB

的信息，不仅给出多达几十亿的网址链接，而且为其中的每一个网站进行了排名。可以说，Alexa 是当前拥有 URL 数量最庞大、排名信息发布最详尽的网站。

③ Cortana（中文名：微软小娜）是微软公司发布的全球第一款个人智能助理。它能够了解用户的喜好和习惯，以及帮助用户进行日程安排、问题回答等。

## Listening Comprehension：Thinking Machines

在线音频

Listen to the article and answer the following 3 questions based on it. After you hear a question, there will be a break of 15 seconds. During the break, you will decide which one is the best answer among the four choices marked (A), (B), (C) and (D).

**Questions**

1. Which of the following is the best for digit numbers calculations? (　　)
   (A) A human
   (B) A cat
   (C) A computer
   (D) None of the above

2. Which of the following is not right? (　　)
   (A) Computers may have trouble understanding a simple conversation.
   (B) Computers might not be able to distinguish between a table and a chair.
   (C) A human might point out the cat in the picture which includes a cat easily.
   (D) A human is better than a computer for digit numbers calculations.

3. Which of the following is right? (　　)
   (A) Humans are still struggling with ways to perform human-like reasoning using a computer.
   (B) Computers are good at computation.
   (C) Computers are good at things that require intelligence.
   (D) All of the above

 **Words**

identification[aiˌdentifiˈkeiʃn] n. 识别，身份证明

 **Phrases**

stand up to　经得起，抵抗
be adept at　擅长，精湛纯熟

## Dictation：Intelligent Agent

在线音频

This article will be played three times. Listen carefully, and fill in the numbered spaces with the appropriate words you have heard.

In artificial intelligence, an Intelligent Agent (IA) is an ___1___ entity which observes through sensors and **acts upon** an environment using **actuators** (i.e. it is an agent) and ___2___ its activity towards achieving goals (i.e. it is "rational", as defined in economics). Intelligent agents may also learn or use knowledge to ___3___ their goals. They may be very simple or very complex. A **reflex machine**, such as a thermostat, is ___4___ an example of an intelligent agent.

Intelligent agents are often described **schematically** as an ___5___ functional system similar to a computer program. For this reason, intelligent agents are sometimes ___6___ abstract intelligent agents (AIA) to distinguish them from their real world ___7___ as computer systems, biological systems, or organizations. Some definitions of intelligent agents ___8___ their autonomy, and so prefer the term autonomous intelligent agents. Still others considered goal-directed ___9___ as the essence of intelligence and so prefer a term ___10___ from economics, "___11___ agent".

Intelligent agents in artificial intelligence are closely ___12___ to agents in economics, and versions of the intelligent agent **paradigm** are ___13___ in cognitive science, ethics, the philosophy of ___14___ reason, as well as in many interdisciplinary socio-cognitive ___15___ and computer social simulations.

Intelligent agents are also ___16___ related to software agents (an autonomous computer program that carries out ___17___ on behalf of users). In computer science, the term intelligent agent may be used to ___18___ to a software agent that has some intelligence. For example, autonomous programs used for ___19___ assistance or data ___20___ (sometimes referred to as bots) are also called "intelligent agents".

## Words

actuator[ˈæktjueitə] n. 效应器
schematically[skiːˈmætikli] adv. 示意性地,图解地,计划性地
paradigm[ˈpærədaim] n. 范式,范例

## Phrases

act upon    作用于
reflex machine    反射机

# Unit 2

## Knowledge Representation and Reasoning

# Part 1

# Reading & Translating

## Section A: Representing and Manipulating Knowledge

Regarding perception we see that understanding images requires a significant amount of knowledge about the items in the image and that the meaning of a sentence might depend on its context (Figure 2-1). These are examples of the role played by the warehouse of knowledge, often called real-world knowledge, maintained by human minds. Somehow, humans store massive amounts of information and draw from that information with remarkable efficiency. Giving machines this capability is a major challenge in artificial intelligence.

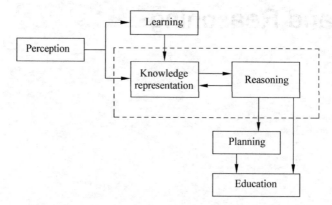

Figure 2-1　Knowledge representation in artificial intelligence

The **underlying** goal is to find ways to represent and store knowledge. This is complicated by the fact that, as we have already seen, knowledge occurs in both **declarative** and procedural forms. Thus, representing knowledge is not merely the representation of facts, but instead encompasses a much broader **spectrum**. Whether a single **scheme** for representing all forms of knowledge will ultimately be found is therefore questionable.

The problem, however, is not just to represent and store knowledge. The knowledge must also be readily accessible, and achieving this accessibility is a challenge. **Semantic** nets are often used as a means of knowledge representation and storage, but extracting information from them can be problematic. For example, the significance of the statement "Mary hit John" depends on the relative ages of Mary and John. (Are the ages 2 and 30 or vice versa?) This information would be stored in the complete semantic net

## Unit 2　Knowledge Representation and Reasoning

but extracting such information during **contextual** analysis could require a significant amount of searching through the net.

Yet another problem dealing with accessing knowledge is identifying knowledge that is **implicitly**, instead of **explicitly**, related to the task **at hand**. Rather than answering the question "Did Arthur win the race?" with a **blunt** "No," we want a system that might answer with "No, he **came down with** the flu and was not able to compete." However, the task is not merely to retrieve related information. We need systems that can distinguish between related information and relevant information. For example, an answer such as "No, he was born in January and his sister's name is Lisa" would not be considered a worthy response to the previous question, even though the information reported is in some way related.

Another approach to developing better knowledge extraction systems has been to insert various forms of reasoning into the extraction process, resulting in what is called **meta-reasoning**—meaning reasoning about reasoning. An example, originally used in the context of database searches, is to apply the closed-world assumption, which is the assumption that a statement is false unless it can be explicitly derived from the information available. For example, it is the closed-world assumption that allows a database to conclude that Nicole Smith does not subscribe to a particular magazine even though the database does not contain any information at all about Nicole. The process is to observe that Nicole Smith is not on the subscription list and then apply the closed-world assumption to conclude that Nicole Smith does not subscribe.

**On the surface** the closed-world assumption appears trivial, but it has consequences that demonstrate how apparently innocent meta-reasoning techniques can have subtle, undesirable effects. Suppose, for example, that the only knowledge we have is the single statement.

　　Mickey is a mouse OR Donald is a duck.

From this statement alone we cannot conclude that Mickey is in fact a mouse. Thus the closed-world assumption forces us to conclude that the statement

　　Mickey is a mouse.

is false. In a similar manner, the closed-world assumption forces us to conclude that the statement

　　Donald is a duck.

is false.

Thus, the closed-world assumption has led us to the contradictory conclusion that although at least one of the statements must be true, both are false. Understanding the consequences of such **innocent-looking** meta-reasoning techniques is a goal of research in the fields of both artificial intelligence and database, and it also **underlines** the complexities involved in the development of intelligent systems.

Finally, there is the problem, known as the **frame** problem, of keeping stored knowledge up to date in a changing environment. If an intelligent agent is going to use its knowledge to determine its behavior, then that knowledge must be current. But the amount of knowledge required to support intelligent behavior can be enormous, and maintaining that knowledge in a changing environment can be a massive **undertaking**. A complicating factor is that changes in an environment often alter other items of information indirectly and accounting for such indirect consequences is difficult. For example, if a flower vase is **knocked over** and broken, your knowledge of the situation no longer contains the fact that water is in the vase, even though spilling the water was only indirectly involved with breaking the vase. Thus, to solve the frame problem not only requires the ability to store and retrieve massive amounts of information in an efficient manner, but it also demands that the storage system properly react to indirect consequences.

## Words

underlying[ˌʌndəˈlaiiŋ] *adj.* 根本的
declarative[diˈklærətiv] *adj.* 说明的，陈述的
spectrum[ˈspektrəm] *n.* 范围，领域
scheme[skiːm] *n.* 计划
semantic[siˈmæntik] *adj.* 语义的
contextual[kənˈtekstʃuəl] *adj.* 上下文的
implicit[imˈplisit] *adj.* 含蓄的，暗中的

explicit[ikˈsplisit] *adj.* 明确的，明白的
blunt[blʌnt] *adj.* 生硬的，直率的
meta-reasoning 元推理
innocent-looking 看上去无恶意的，看上去无害的
underline[ˌʌndəˈlain] *v.* 强调
frame[freim] *n.* 框架
undertaking[ˌʌndəˈteikiŋ] *n.* 任务，事业

## Phrases

at hand　　在手边，在附近
come down with　　染上病
on the surface　　在表面上，外表上
knock over　　打翻，撞倒

## Exercises

I. Read the following statements carefully, and decide whether they are true (T) or false (F) according to the text.

　　____ 1. To solve the frame problem not only requires the ability to store and retrieve massive amounts of information in an efficient manner, but it also demands that the storage system properly react to direct consequences.

## Unit 2  Knowledge Representation and Reasoning

____ 2. The closed-world assumption has led us to the contradictory conclusion that although at least one of the statements must be false, both are true.

____ 3. Representing knowledge is not merely the representation of facts, but instead includes a much broader spectrum.

____ 4. Knowledge representation and storage often use semantic nets as a means.

____ 5. Meta-reasoning means reasoning about reasoning.

II. Choose the best answer to each of the following questions according to the text.

1. Which of the following is used as a means by knowledge representation and storage? (　　)
   A. Turing test
   B. Programming
   C. Semantic nets
   D. None of the above

2. Which of the following is not right? (　　)
   A. To solve the frame problem not only requires the ability to store and retrieve massive amounts of information in an efficient manner, but it also demands that the storage system properly react to direct consequences.
   B. Meta-reasoning means reasoning about reasoning.
   C. Knowledge representation and storage often use semantic nets as a means.
   D. All of the above

3. Which of the following is right? (　　)
   A. The closed-world assumption has led us to the contradictory conclusion that although at least one of the statements must be false, both are true.
   B. If an intelligent agent is going to use its knowledge to determine its behavior, then that knowledge must not be current.
   C. The amount of knowledge required to support intelligent behavior can be enormous, and maintaining that knowledge in a changing environment can be a massive undertaking.
   D. All of the above

III. Fill in the numbered spaces with the words or phrases chosen from the box. Change the forms where necessary.

> argument   infer   predicate   between   define
> combine   purpose   break   refer   relationship

**Predicate Logic**

　　In propositional logic, a symbol that represents a sentence is atomic: it cannot be __1__ up to find information about its components. For example, consider the sentences:

P1: "Linda is Mary's mother"　　P2: "Mary is Anne's mother"

We can ___2___ these two sentences in many ways to create other sentences, but we cannot extract any relation between Linda and Anne. For example, we cannot ___3___ from the above two sentences that Linda is the grandmother of Anne. To do so, we need predicate logic: the logic that ___4___ the relation between the parts in a proposition.

In predicate logic, a sentence is divided into a predicate and ___5___. For example, each of the following propositions can be written as ___6___ with two arguments:

P1: "Linda is Mary's mother" becomes mother (Linda, Mary)

P2: "Mary is Anne's mother" becomes mother (Mary, Anne)

The ___7___ of motherhood in each of the above sentences is defined by the predicate mother. If the object Mary in both sentences ___8___ to the same person, we can infer a new relation ___9___ Linda and Anne: grandmother (Linda, Anne). This is the whole ___10___ of predicate logic.

**IV. Translate the following passage into Chinese.**

**Reasoning and Logic**

Reasoning is the action of constructing thoughts into a valid argument. This is something you probably do every day. When you make a decision, you are using reasoning, taking different thoughts and making those thoughts into reasons why you should go with one option over the other options available. When you construct an argument, that argument will be either valid or invalid. A valid argument is reasoning that is comprehensive on the foundation of logic or fact.

Inductive and deductive reasoning are both forms of propositional logic. Propositional logic is the branch of logic that studies ways of joining and/or modifying entire propositions, statements or sentences to form more complicated propositions, statements or sentences. Inductive and deductive reasoning use propositional logic to develop valid arguments based on fact and reasoning. Both types of reasoning have a premise and a conclusion. How each type of reasoning gets to the conclusion is different.

## Section B: Learning

In addition to representing and manipulating knowledge, we would like to give intelligent agents (Figure 2-2) the ability to acquire new knowledge. We can always "teach" a computer-based agent by writing and installing a new program or explicitly adding to its stored data, but we would like intelligent agents to be able to learn on their own. We want agents to adapt to changing environments and to perform tasks for which we cannot easily write programs in advance. A robot designed for household **chores** will be faced with new furniture, new appliances, new pets, and even new owners. An autonomous, self-driving car must adapt to variations in the boundary lines on roads. Game-playing agents should be able to develop and apply new strategies.

## Unit 2  Knowledge Representation and Reasoning

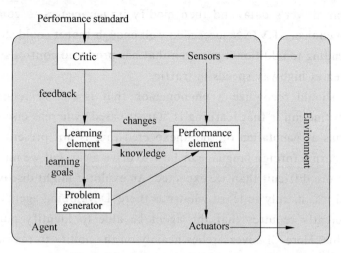

Figure 2-2  Agent in artificial intelligence

One way of classifying approaches to computer learning is by the level of human intervention required. At the first level is learning by imitation, in which a person directly demonstrates the steps in a task (perhaps by carrying out a sequence of computer operations or by physically moving a robot through a sequence of motions) and the computer simply records the steps. This form of learning has been used for years in application programs such as spreadsheets and word processors, where frequently occurring sequences of commands are recorded and later replayed by a single request. Note that learning by imitation places little responsibility on the agent.

At the next level is learning by supervised training. In supervised training a person identifies the correct response for a series of examples and then the agent generalizes from those examples to develop an algorithm that applies to new cases. The series of examples is called the training set. Typical applications of supervised training include learning to recognize a person's handwriting or voice, learning to distinguish between junk and welcome email, and learning how to identify a disease from a set of symptoms.

A third level is learning by reinforcement. In learning by reinforcement, the agent is given a general rule to judge for itself when it has succeeded or failed at a task during trial and error. Learning by reinforcement is good for learning how to play a game like chess or **checkers**, because success or failure is easy to define. In contrast to supervised training, learning by reinforcement allows the agent to act autonomously as it learns to improve its behavior over time.

Learning remains a challenging field of research since no general, universal principle has been found that covers all possible learning activities. However, there are numerous examples of progress. One is ALVINN (Autonomous Land Vehicle in a Neural Net), a system developed at Carnegie Mellon University to learn to **steer** a van with **an on-board** computer using a video camera for input. The approach used was supervised training. ALVINN collected data from a human driver and used that data to adjust its own steering decisions. As it learned, it would predict where to steer, check its prediction against the human driver's data, and then modify its **parameters** to come closer to the human's steering choice. ALVINN succeeded well enough that it could steer the van at 70 miles an hour, leading to additional research that has produced control systems that have successfully driven at highway speeds in traffic.

Finally, we should recognize a phenomenon that is closely related to learning: discovery. The distinction is that learning is "target based" whereas discovery is not. The term discovery has a **connotation** of the unexpected that is not present in learning. We might **set out to** learn a foreign language or how to drive a car, but we might discover that those tasks are more difficult than we expected. An explorer might discover a large lake, whereas the goal was merely to learn what was there. Developing agents with the ability to discover efficiently requires that the agent be able to identify potentially **fruitful** "trains of thought." Here, discovery relies heavily on the ability to reason and the use of heuristics. Moreover, many potential applications of discovery require that an agent be able to distinguish meaningful results from insignificant ones. A data mining agent, for example, should not report every trivial relationship it finds.

Examples of success in computer discovery systems include Bacon, named after the philosopher Sir Francis Bacon, that has discovered (or maybe we should say "rediscovered") Ohm's law of electricity, Kepler's third law of **planetary** motion, and the conservation of **momentum**. Perhaps more **persuasive** is the system AUTOCLASS that, using infrared **spectral** data, has discovered new classes of stars that were previously unknown in astronomy—a true scientific discovery by a computer.

## Words

chore[tʃɔː(r)] *n.* 家庭杂务
checker['tʃekə(r)] *n.* 国际跳棋
steer[i'rædikeit] *v.* 驾驶(船、汽车等)
on-board  在船(或飞机、车)上的
parameter[pə'ræmitə(r)] *n.* 参数
connotation[ˌkɒnə'teiʃn] *n.* 内涵

fruitful['fruːtfl] *adj.* 富有成效的
planetary['plænətri] *adj.* 行星的
momentum[mə'mentəm] *n.* 动量,动力
persuasive[pə'sweisiv] *adj.* 有说服力的
spectral['spektrəl] *adj.* 光谱的

## Phrases

set out to   打算,着手

## Exercises

I. Read the following statements carefully, and decide whether they are true (T) or false (F) according to the text.

_____ 1. ALVINN is a system developed at Washington University.

# Unit 2  Knowledge Representation and Reasoning

____ 2. Learning by imitation is the same as learning by supervised training.

____ 3. Supervised training allows the agent to act autonomously as it learns to improve its behavior over time.

____ 4. Typical applications of supervised training include learning how to identify a disease from a set of symptoms.

____ 5. Learning by imitation places little responsibility on the agent.

II. **Choose the best answer to each of the following questions according to the text.**

1. Which of the following is right about ALVINN? (    )
   A. ALVINN succeeded well enough that it could steer the van at 70 miles.
   B. ALVINN succeeded well enough that it could steer the van at 80 miles.
   C. ALVINN succeeded well enough that it could steer the van at 90 miles.
   D. ALVINN succeeded well enough that it could steer the van at 100 miles.

2. Which of the following is mentioned about learning by reinforcement? (    )
   A. Learning by reinforcement has been used for years in application programs such as spreadsheets and word processors.
   B. Learning by reinforcement allows the agent to act autonomously as it learns to improve its behavior over time.
   C. Learning by reinforcement places little responsibility on the agent.
   D. None of the above

3. Which of the following is right? (    )
   A. Discovery is "target based" whereas learning is not.
   B. The term learning has a connotation of the unexpected that is not present in discovery.
   C. Many potential applications of discovery require that an agent be able to distinguish meaningful results from insignificant ones.
   D. All of the above

III. **Fill in the numbered spaces with the words or phrases chosen from the box. Change the forms where necessary.**

> accept   practice   call   equivalent   set
> though   likely   opposite   refer   express

**Deductive and Inductive Reasoning**

　　The two major types of reasoning, deductive and inductive, ___1___ to the process by which someone creates a conclusion as well as how they believe their conclusion to be true.

　　Deductive reasoning requires one to start with a few general ideas, ___2___ premises, and apply them to a specific situation. Recognized rules, laws, theories, and other widely _____3_____ truths are used to prove that a conclusion is right. The concept of deductive reasoning is often _____4_____ visually using a funnel that narrows a general idea into a specific conclusion. In _____5_____, the most basic form of deductive reasoning is the syllogism, where two premises that share some idea support a conclusion.

Inductive reasoning uses a _____6_____ of specific observations to reach an overarching conclusion; it is the _____7_____ of deductive reasoning. So, a few particular premises create a pattern which gives way to a broad idea that is _____8_____ true. This is commonly shown using an inverted funnel (or a pyramid) that starts at the narrow premises and expands into a wider conclusion. There is no _____9_____ to a syllogism in inductive reasoning, meaning there is no basic standard format. All forms of inductive reasoning, _____10_____, are based on finding a conclusion that is most likely to fit the premises and is used when making predictions, creating generalizations, and analyzing cause and effect.

IV. Translate the following passage into Chinese.

**Semantic Network**

Semantic network or semantic net was proposed by Quillian in 1967 in order to represent the knowledge in a form of graph. Semantic network is a technique of knowledge representation that is used for propositional information, and sometimes called a propositional net. In knowledge representation the semantic networks are two dimensional. In terms of mathematics a semantic network is defined as a labeled directed graph. The semantic network is composed of links, nodes and link labels. In the diagram the semantic network nodes are described as ellipses, circles or rectangles to show objects such as physical objects, situations or concepts. The links can be used to express the relationships between objects. A particular relation is specified by link labels. The basic structure of knowledge organizing is provided by relationships.

# Part 2

# Simulated Writing: Uncovering the Secrets of Clear Writing（II）

接15页

**3. 说明数据**

当需要对比数据时，可以使用图表来说明。使用图形来表示数据可以使信息更易于理解和记忆。图形可以使数字信息有意义，可揭示其背后的趋势和模式，简化复杂的关系，并为文档增加视觉吸引力。要熟悉最流行的图表类型、它们的优缺点，以及什么时候使用它们才合适。表2-1列出了在企业文档中使用图表的注意事项。

## Unit 2  Knowledge Representation and Reasoning

表 2-1  使用图表的注意事项

| 图表类型 | 适合提到 | 尽量避免 |
| --- | --- | --- |
| 条形图和柱状图 | • 展示随时间变化的数据<br>• 比较项目<br>• 在比较持续时间时选择柱状图 | • 比较过多的项目，5 个或 6 个是比较典型的上限<br>• 当类别有很长的名字时不要使用条形图；而是使用柱状图 |
| 线形图 | • 展示数据的趋势或模式 | • 如果数值被均匀分割，使用线形图，例如按月或年划分 |
| 饼图 | • 展示一类数据的大小相对于其他类和整体的比例 | • 在饼图中展示多于一个类别的数据 |
| 过程图 | • 展示过程中的步骤<br>• 使用常见的形状来表示过程中的部分 | • 改变常见形状的意义 |
| 层次图 | • 展示一个组织中的汇报关系 | • 在图内的每个方框中包含过多的细节 |

1）条形图和柱状图

条形图和柱状图表示以间距归类的分类数据和数字数据，例如每月的销售额或每个产品的费用。条形图的每一个类别都包含一个横向的条形，并且，每一个条形的高度或长度代表着那一类别的值。柱状图与其相似，只是使用纵向的条形。图 2-3 展示了柱状图的实例。

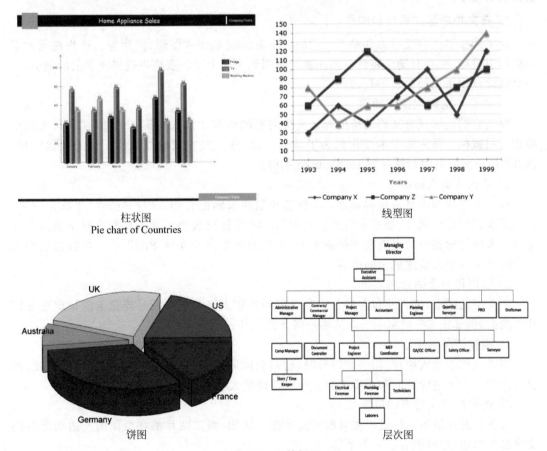

柱状图
Pie chart of Countries

线型图

饼图

层次图

图 2-3  图的例子

2) 线形图

线形图揭示了数据的趋势和模式。线形图展示了两个数值是如何彼此关联的。纵轴(y)通常表示数量,例如金额或者百分比。横轴(x)通常表示时间单元。因此,用线形图展示随时间变化的数量最为理想。图 2-3 同样展示了线形图。

3) 饼图

饼图可以被分割成若干个楔形块,每块都代表一个类别。有时,为了特别强调,将一个楔形块从饼图中分离出来。饼图可以将整体和其各部分进行对比。图 2-3 同样展示了带有重点强调的楔形块的饼图。

4) 过程图

过程图展示了一个过程中的若干步骤,有时称作流程图。不同的形状代表各种不同类型的活动。例如,圆形或椭圆代表过程的开始和结束,菱形代表必须进行的决策和选择,而矩形代表过程的主要活动或者步骤。

5) 层次图(或组织结构图)

当人们和物体按照一定层次组织在一起时,可以使用一个层次图来代表它们。层次图通常画为一个水平的或竖直的树,使用几何图形来代表其不同的元素。线将各种形状连接起来表明元素之间的关系。

层次图展示了企业的正式结构。层次图展示了对象间的关系。图 2-3 同样展示了一个层次图的例子。

**4. 在文档中添加表格和图片**

让文档变得更有吸引力和易于阅读的一个方法就是插入图形、照片等。这些视觉元素可以吸引眼球,并且有助于将读者的注意力吸引到文字上来。表格也提供视觉上的吸引力,而且被设计用来比较信息的列表。

1) 在合适的时候使用插图

仅当它们能够提升文档的价值并且支持内容的时候才插入图形。例如,在公司文档中使用公司徽标。插入图形不要仅仅为了装饰页面、分割文本或增加文档长度的视觉资料。图 2-4 展示了带有附加图片的 InfoSource 手册。

2) 给图片加上标签

每一个图片和表格通常都包含一个标签和题注来标记它们。使用"图♯"来标记图片,比如图表、示意图、照片、地图和绘图。使用"表♯"来标记表格。给图片和表格单独按顺序编号。表格的标签通常出现在表格的上方,图片的标签出现在插图的下方。在标签后面加一个简短的说明来描述展示的内容。

3) 引用每一个图片或表格

在文字的附近,插入所包含的每一个引用图片和表格。参考图号理论上应该出现在图片的前面,例如图 2-5 展示了云端的客户服务的照片。

4) 调整插图的大小

当在文档中插入照片、图表和图形时,它们会以原始大小和分辨率显示。如果必要,调整图像的大小来适应页面和平衡内容。确保在调整大小的时候保持纵横比不变。

5) 将插图放在适当的位置

将图片放在适当的位置以使页面布局平衡。例如,两张图片都在页面的左边而所有的文字都在右边,这可能看起来不平衡。

# Unit 2　Knowledge Representation and Reasoning

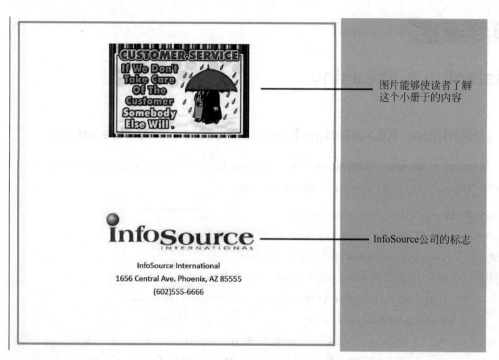

图 2-4　添加到 InfoSource 小册子的图片

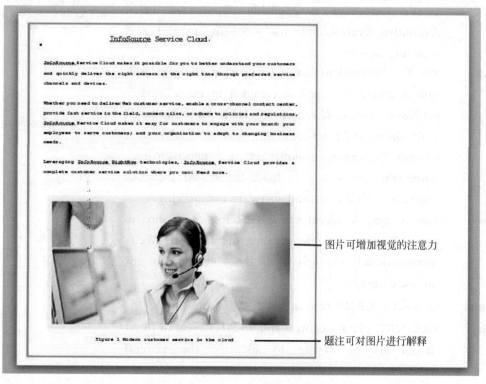

图 2-5　添加小册子中的图片和标题

# Part 3

# Listening & Speaking

## Dialogue: Knowledge Representation and Reasoning

(Today is Monday. Henry and Mark are on the way to the classroom when they **come across** Sophie who is going to the classroom too.)

**Henry & Mark:** Good morning, Sophie.

**Sophie:** Good morning, Henry and Mark.

**Henry:** How's your weekend going, Sophie?

**Sophie:** Not bad. I'm learning knowledge representation and reasoning. It is not very easy to understand. Do you know anything about it?

**Henry:** As I know,[1] Knowledge Representation and Reasoning (KRR) is an exciting, **well-established** field of research. In KRR, a fundamental assumption is that an agent's knowledge is explicitly represented in a declarative form, suitable for processing by dedicated reasoning engines.

**Mark:** Yes. This assumption, that much of what an agent deals with is knowledge-based, is common in many modern intelligent systems. Consequently, KRR has contributed to the theory and practice of various areas in AI, such as automated planning, natural language understanding, **among others**, as well as to fields beyond AI, including databases, verification, and software engineering.

**Henry:** That is right. In recent years KRR has contributed to new and emerging fields including the semantic Web, computational biology, and the development of software agents.

**Sophie:** So what is KRR's role in AI?

**Henry:** Well, KRR plays a central role in AI. Research in AI **started off** by trying to identify the general mechanisms responsible for intelligent behavior.

**Mark:** However, it quickly became obvious that general and powerful methods are not enough to get the desired

[1] Replace with:
1. As far as I know,
2. So far as I know,
3. As for as I know,

## Unit 2 Knowledge Representation and Reasoning

result, namely, intelligent behavior. Almost all tasks a human can perform which are considered to require intelligence are also based on a huge amount of knowledge. For instance, understanding and producing natural language relies heavily on knowledge about the language, about the structure of the world, about social relationships, etc.

**Sophie:** A little bit abstract. Could you please detail KRR?

**Henry:** Sure. Knowledge representation and reasoning are closely **coupled** components; Each is **intrinsically** tied to the other. A representation scheme is not meaningful on its own; It must be useful and helpful in achieving certain tasks. The same information may be represented in many different ways, depending on how you want to use that information.

**Mark:** Let me give you an example. In mathematics, if we want to solve problems about ratios, we would most likely use algebra, but we could also use simple hand drawn symbols. To say half of something, you could use 0.5x or you could draw a picture of the object with half of it colored differently. Both would convey the same information but the former is more compact and useful in complex scenarios where you want to perform reasoning on the information. It is important at this point to understand how knowledge representation and reasoning are interdependent components, and as AI system designer, you have to consider this relationship when **coming up with** any solution.

**Sophie:** And any others about KRR?

**Henry:** Well, the key question when we begin to think about knowledge representation and reasoning is how to **approach** the problem—should we try to emulate the human brain completely and exactly as it is? Or should we come up with something new?

**Mark:** Additionally,[2] since we do not know how the knowledge representation and reasoning components are implemented in humans, even though we can see their **manifestation** in the form of intelligent behavior, we need a synthetic (artificial) way to model

[2] Replace with:
1. In addition,
2. Moreover,
3. Besides,

035

|  |  |
|---|---|
|  | the knowledge representation and reasoning capability of humans in computers. |
| Sophie: | Well, from your description, I'm sure knowledge representation and reasoning will be applied more in AI. |
| Henry: | I think so. |
| Mark: | I'm pretty confident. |

 **Exercises**

Work in a group, and make up a similar conversation by replacing the statements with other expressions on the right side.

 **Words**

well-established 得到确认的
couple ['kʌpl] v. 结合,连接
intrinsically [ ɪn'trɪnzɪkli; ɪn'trɪnsɪkli ]
    adj. 内在地,本质地,固有地

approach [ə'prəʊtʃ] v. 处理
manifestation [ˌmænɪfe'steɪʃn] n. 表现,显示

 **Phrases**

come across    偶然遇见
among others    除了别的之外,尤其
start off    开始
come up with    提出,想出

在线音频

## Listening Comprehension: Logical Reasoning

*Listen to the article and answer the following 3 questions based on it. After you hear a question, there will be a break of 15 seconds. During the break, you will decide which one is the best answer among the four choices marked (A), (B), (C) and (D).*

**Questions**

1. Usually how many major types is logical reasoning divided into? (     )

    (A) One

    (B) Two

    (C) Three

    (D) Four

# Unit 2　Knowledge Representation and Reasoning

2. Which of the following is right? (　　)

    (A) It generally forms the bulk of most logic-based arguments although inductive logical reasoning can be far more complex to understand than deductive reasoning.

    (B) Given that the premises are accurate, deductive reasoning can prove an absolute truth or fact.

    (C) Given that the premises are accurate, inductive logic uses premises to determine a highly probable, but not absolute, conclusion.

    (D) All of the above

3. Which of the following must first be presented to reach a conclusion using logical reasoning? (　　)

    (A) Conclusion

    (B) Principle

    (C) Fact

    (D) None of the above

## Words

system [ˈsistəm] n. 方法
premise [ˈpremis] n. 前提,假设
deductive [diˈdʌktiv] adj. 演绎的,推论的,推断的
inductive [inˈdʌktiv] adj. 归纳的
argument [ˈɑːgjumənt] n. 论证,论据
fallacy [ˈfæləsi] n. 谬论,谬误

grocer [ˈgrəusə(r)] n. 杂货店,食品商
beet [biːt] n. 甜菜
turnip [ˈtɜːnip] n. 萝卜
shipment [ˈʃipmənt] n. 装载的货物
indisputable [ˌindiˈspjuːtəbl] adj. 不容置疑的,无可争辩的
bulk [bʌlk] n. 大多数,大部分

## Phrases

a set of　　一组
break down into　　分解成……
add up to　　合计达,总计达

## Dictation: Semantic Networks

*This article will be played three times. Listen carefully, and fill in the numbered spaces with the appropriate words you have heard.*

Semantic networks were ＿＿1＿＿ in the early 1960s by Richard H. Richens. A semantic network uses ＿＿2＿＿ graphs to represent knowledge. A directed graph is

在线音频

037

made of **vertices**（nodes）and \_\_\_\_3\_\_\_\_（arcs）. Semantic networks use vertices to represent \_\_\_\_4\_\_\_\_, and edges（**denoted** by arrows）to represent the relation between two concepts.

To develop an \_\_\_\_5\_\_\_\_ definition of a concept, experts have \_\_\_\_6\_\_\_\_ the definition of concepts to the theory of \_\_\_\_7\_\_\_\_. A concept, \_\_\_\_8\_\_\_\_, can be thought of as a set or a subset. For example, \_\_\_\_9\_\_\_\_ defines the set of all animals; \_\_\_\_10\_\_\_\_ defines the set of all horses and is a subset of the set animal. An object is a member (**instance**) of a set. Concepts are \_\_\_\_11\_\_\_\_ by vertices.

In a semantic network, relations are shown by edges. An edge can \_\_\_\_12\_\_\_\_ a subclass relation—the edge is directed from the \_\_\_\_13\_\_\_\_ to its superclass. An edge can also define an \_\_\_\_14\_\_\_\_ relation—the edge is directed from the instance to the set to which it belongs. An edge can also define an \_\_\_\_15\_\_\_\_ of an object (color, size, …). Finally, an edge can define a **property** of an \_\_\_\_16\_\_\_\_, such as possessing another object. One of the most important relations that can be well defined in a semantic network is \_\_\_\_17\_\_\_\_. An inheritance relation defines the \_\_\_\_18\_\_\_\_ that all the attributes of a class are \_\_\_\_19\_\_\_\_ in an inherited class. This can be used to \_\_\_\_20\_\_\_\_ new knowledge from the knowledge represented by the graph.

## Words

vertices['vɜːtisiːz] *n*. 顶点，至高点（vertex 的复数）
denote[di'nəut] *v*. 表示，指示
instance['instəns] *n*. 实例
property['prɔpəti] *n*. 财产，所有权

# Unit 3

# Reasoning with Uncertainty

# Part 1

# Reading & Translating

## Section A: Reasoning with Uncertainty

### 1. Introduction

Though there are various types of uncertainty in various aspects of a reasoning system, the "reasoning with uncertainty" (or "reasoning under uncertainty") research in AI has been focused on the uncertainty of truth value, that is, to allow and process truth values other than "true" and "false".

Generally speaking, to develop a system that reasons with uncertainty means to provide the following:

- a semantic explanation about the origin and nature of the uncertainty
- a way to represent uncertainty in a **formal language**
- a set of inference rules that derive uncertain (though **well-justified**) conclusions
- an efficient memory-control mechanism for uncertainty management

### 2. Non-monotonic logics

A reasoning system is monotonic if the truthfulness of a conclusion does not change when new information is added to the system—the set of **theorem** can only monotonically grows when new **axioms** are added. In contrast, in a system doing non-monotonic reasoning the set of conclusions may either grow or shrink when new information is obtained.

Non-monotonic logics are used to formalize plausible reasoning, such as the following inference step:

Birds typically fly.
Tweety is a bird.
---------------------------
Tweety (presumably) flies.

Such reasoning is characteristic of **commonsense** reasoning, where default rules are applied when case-specific information is not available.

The conclusion of non-monotonic argument may turn out to be wrong. For example, if Tweety is a penguin, it is incorrect to conclude that Tweety flies. Non-monotonic reasoning often requires jumping to a conclusion and **subsequently retracting** that conclusion as further information becomes available.

All systems of non-monotonic reasoning **are concerned with** the issue of consistency.

Inconsistency is resolved by removing the relevant conclusion(s) derived previously by default rules. Simply speaking, the truth value of **propositions** in a non-monotonic logic can be classified into the following types:
- facts that are definitely true, such as "Tweety is a bird"
- default rules that are normally true, such as "Birds fly"
- **tentative** conclusions that are presumably true, such as "Tweety flies"

When an inconsistency is recognized, only the truth value of the last type is changed.

Revising a knowledge base often follows the principle of minimal change: one conserves as much information as possible.

One approach towards this problem is truth maintenance system, in which a "justification" for each proposition is stored, so that when some propositions are rejected, some others may need to be removed, too.

Major problems in these approaches:
- conflicts in defaults, such as in the "Nixon Diamond" [1]
- computational expense: to maintain the consistency in a huge knowledge base is hard, if not impossible

### 3. Probabilistic reasoning

Basic idea: to use probability theory to represent and process uncertainty. In probabilistic reasoning, the truth value of a proposition is extended from $\{0,1\}$ to $[0,1]$, with binary logic as its special case.

**Justification**: though no conclusion is absolutely true, the one with the highest probability is preferred. Under certain assumptions, probability theory gives the optimum solutions.

To extend the basic Boolean connectives to probability functions:
- **negation**: $P(\neg A) = 1 - P(A)$
- **conjunction**: $P(A \wedge B) = P(A) * P(B)$ if A and B are independent of each other
- **disjunction**: $P(A \vee B) = P(A) + P(B)$ if A and B never happen at the same time

Furthermore, the conditional probability of B given A is $P(B|A) = P(B \wedge A)/P(A)$, from which Bayes' Theorem is derived, and it is often used to update a system's belief according to new information: $P(H|E) = P(E|H) * P(H)/P(E)$.

Bayesian Networks are directed **acyclic graphs** in which the nodes represent variables of interest and the links represent informational or causal dependencies among the variables. The strength of dependency is represented by conditional probabilities. Compared to other approaches of probabilistic reasoning, Bayesian network (Figure 3-1) is more efficient, though its actual computational cost is still high for complicated problems.

Challenges to probabilistic approaches:
- unknown probability values

## Bayesian Network Motivation

- We want a representation and reasoning system that is based on conditional (and marginal) independence
  - Compact yet expressive representation
  - Efficient reasoning procedures
- Bayesian(Belief) Networks are such a representation
  - Named after thomas Bayes(ca. 1702—1761)
  - Term coined in 1985 by Judea Pearl(1936—)
  - Their invention changed the primary focus of AI from logic to probability

Thomas Bayes

Judea Pearl

Pearl just received the ACM Turing Award (widely considered the "Nobel Prize in Computing") for his contributions to Artificial Intelligence!

Figure 3-1　Bayesian Network

- inconsistent probability assignments
- computational expense

Considering the uncertainty in probability judgments, some people go further to study imprecise probability.

**4. Fuzzy logic**

Fuzzy logic is a generalization of classical logic, and reflects the imprecision of human language and reasoning.

Examples of fuzzy concepts: "young", "furniture", "most", "cloudy", and so on.

According to fuzzy logic, whether an instance belongs to a concept is usually not a matter of "yes/no", but a matter of degree. Fuzzy logic uses a degree of membership, which is a real number in $[0,1]$.

A major difference between this number and probability is: the uncertainty in fuzzy concepts usually does not get reduced with the coming of new information. Compare the following two cases:

- I'm afraid that tomorrow will be cloudy, so let's take the picture today.
- I'm not sure whether the current weather should be classified as "cloudy" or not.

Basic fuzzy operators:

- negation: $M(\neg A) = 1 - M(A)$.
- conjunction: $M(A \wedge B) = \min\{M(A), M(B)\}$.
- disjunction: $M(A \vee B) = \max\{M(A), M(B)\}$.

Typically, in building a fuzzy system, the designer needs to provide all membership functions included in it, by considering how the concepts are used by **average people**. Most successful applications of fuzzy logic so far are in fuzzy control systems, where expert knowledge is coded into fuzzy rules.

Challenges to fuzzy approaches:

- degree of membership is often context dependent

- general-purpose fuzzy rules are hard to get

5. **Truth-value as evidential support**

This approach is taken in the **NARS** project, an intelligent reasoning system.

The basic idea is to see the truth-value of a statement as measuring the evidential support the statement gets from the system's experience. Such a truth-value consists of two factors: frequency (the proportion of positive evidence among available evidence) and confidence (the proportion of currently available evidence among all evidence at a near future).

This approach attempts to uniformly represent various types of uncertainty.

## Words

| | |
|---|---|
| well-justified 合理的<br>monotonic [ˈmɒnə(u)ˈtɒnɪk] adj. 单调的，无变化的<br>theorem [ˈθɪərəm] n. 定理，原理<br>axiom [ˈæksɪəm] n. 公理<br>commonsense [ˈkɒmənˈsens] adj. 具有常识的，常识的<br>subsequently [ˈsʌbsɪkwəntlɪ] adv. 随后，其后 | retract [rɪˈtrækt] v. 取消，收回，缩回<br>proposition [ˌprɒpəˈzɪʃn] n. 命题<br>tentative [ˈtentətɪv] adj. 试验性的，暂定的<br>justification [ˌdʒʌstɪfɪˈkeɪʃn] n. 理由，认为有理<br>negation [nɪˈgeɪʃn] n. 否定，否认<br>conjunction [kənˈdʒʌŋkʃn] n. 连词，结合<br>disjunction [dɪsˈdʒʌŋkʃn] n. 析取，分离 |

## Phrases

formal language 形式语言
be concerned with 涉及
acyclic graph 非循环图
average people 普通人，一般人

## Abbreviations

NARS  Non-Axiomatic Reasoning System   非公理推荐系统

## Notes

[1] 尼克松菱形(Nixon Diamond)是包括两个扩展的缺省理论,其缺省逻辑是(美国)共和党人不爱好和平,教友会爱好和平。因为尼克松既是共和党的人又是教友会的人,两个

扩展都可以应用。但是，应用第一个扩展导致尼克松是不爱好和平的人的结论。以同样的方式，应用第二个扩展得出尼克松是爱好和平的人，因此使第一个扩展不可应用。这种特定的缺省理论因此有两个结论，其中一个是真，而另一个是假，就导致了矛盾。

 **Exercises**

I. Read the following statements carefully, and decide whether they are true (T) or false (F) according to the text.

　　____ 1. The conclusion of non-monotonic argument is always true.

　　____ 2. Probabilistic reasoning is more efficient than Bayesian network.

　　____ 3. Conflicts in the "Nixon Diamond" is an example of conflicts in defaults.

　　____ 4. Inconsistent probability assignments is one of the challenges to probabilistic approaches.

　　____ 5. That general-purpose fuzzy rules are hard to get is one of the challenges to fuzzy approaches.

II. Choose the best answer to each of the following questions according to the text.

　1. When was Thomas Bayes born? (　　)

　　A. In 1936

　　B. In 1702

　　C. In 1761

　　D. In 1985

　2. Which of the following is right? (　　)

　　A. According to fuzzy logic, whether an instance belongs to a concept is usually not a matter of "yes/no", but a matter of degree.

　　B. Fuzzy logic uses a degree of membership, which is a real number in $[0,1]$.

　　C. Fuzzy logic is a generalization of classical logic, and reflects the imprecision of human language and reasoning.

　　D. All of the above

　3. Which of the following is not the challenge to probabilistic approaches? (　　)

　　A. Degree of membership.

　　B. Computational expense

　　C. Unknown probability values

　　D. Inconsistent probability assignments

III. Fill in the numbered spaces with the words or phrases chosen from the box. Change the forms where necessary.

　　　　┌─────────────────────────────────────────┐
　　　　│　some　　typical　　kind　　particular　　proposition　│
　　　　│　fuzzy　　base　　introduce　　refer　　aim　　　　　│
　　　　└─────────────────────────────────────────┘

# Unit 3  Reasoning with Uncertainty

**Fuzzy Logic**

Fuzzy logic emerged in the context of the theory of fuzzy sets, ___1___ by Zadeh (1965). A ___2___ set assigns a degree of membership, ___3___ a real number from the interval $([0,1])$, to elements of a universe. Fuzzy logic arises by assigning degrees of truth to propositions. The standard set of truth values (degrees) is $([0,1])$, where $(0)$ represents "totally false", $(1)$ represents "totally true", and the other numbers ___4___ to partial truth, i.e., intermediate degrees of truth.

"Fuzzy logic" is often understood in a very wide sense which includes all ___5___ of formalisms and techniques referring to the systematic handling of degrees of ___6___ kind. In ___7___ in engineering contexts (fuzzy control, fuzzy classification, soft computing) it is ___8___ at efficient computational methods tolerant to suboptimality and imprecision. It focuses on logics ___9___ on a truth-functional account of partial truth and studies them in the spirit of classical mathematical logic (syntax, model theoretic semantics, proof systems, completeness, etc.; both at ___10___ and the predicate level).

**IV. Translate the following passage into Chinese.**

**Non-monotonic Logic**

Everyday reasoning is mostly non-monotonic because it involves risk: we jump to conclusions from deductively insufficient premises. We know when it is worthwhile or even necessary (for example, in medical diagnosis) to take the risk. Yet we are also aware that such inference is "defeasible"—that new information may undermine old conclusions. Various kinds of defeasible but remarkably successful inference have traditionally captured the attention of philosophers (theories of induction, Peirce's theory of abduction, inference to the best explanation, and so on). More recently logicians have begun to approach the phenomenon from a formal point of view. The result is a large body of theories at the interface of philosophy, logic and artificial intelligence.

## Section B: Probabilistic Reasoning

**Probability** theory is used to discuss events, categories, and hypotheses about which there is not 100% certainty.

We might write A→B, which means that if A is true, then B is true. If we are unsure whether A is true, then we cannot make use of this expression.

In many real-world situations, it is very useful to be able to talk about things that lack certainty. For example, what will the weather be like tomorrow? We might formulate a very simple hypothesis based on general observation, such as "it is sunny only 10% of the time, and rainy 70% of the time". We can use a notation similar to that used for predicate calculus to express such statements:

P(S) = 0.1
P(R) = 0.7

The first of these statements says that the probability of S ("it is sunny") is 0.1. The second says that the probability of R is 0.7. Probabilities are always expressed as real numbers between 0 and 1. A probability of 0 means "definitely not" and a probability of 1 means "definitely so." Hence, P(S) = 1 means that it is always sunny.

Many of the operators and notations that are used in prepositional logic can also be used in probabilistic notation. For example, P($\neg$ S) means "the probability that it is not sunny"; P(S $\wedge$ R) means "the probability that it is both sunny and rainy." P(A $\vee$ B), which means "the probability that either A is true or B is true," is defined by the following rule: P(A $\vee$ B) = P(A) + P(B) − P(A $\wedge$ B)

The notation P(B|A) can be read as "the probability of B, given A." This is known as conditional probability—it is conditional on A. In other words, it states the probability that B is true, given that we already know that A is true. P(B|A) is defined by the following rule: Of course, this rule cannot be used in cases where P(A) = 0.

For example, let us suppose that the likelihood that it is both sunny and rainy at the same time is 0.01. Then we can calculate the probability that it is rainy, given that it is sunny as follows:

P(R|S) = P(R $\wedge$ S)/P(S) = 0.01/0.1 = 0.1

The basic approach statistical methods adopt to deal with uncertainty is via the axioms of probability: Probabilities are (real) numbers in the range 0 to 1.

A probability of P(A) = 0 indicates total uncertainty in A, P(A) = 1 total certainty and values in between some degree of (un)certainty.

Probabilities can be calculated in a number of ways.

Probability = (number of desired outcomes)/(total number of outcomes)

So given a pack of playing cards the probability of being dealt an **ace** from a full normal **deck** is 4 (the number of aces)/52 (number of cards in deck) which is 1/13. Similarly the probability of being dealt a **spade suit** is 13/52 = 1/4.

If you have a choice of number of items k from a set of items n then the

$C_n^k = \dfrac{n!}{k!(n-k)!}$ formula is applied to find the number of ways of making this choice. (! = **factorial**).

So the chance of winning the national lottery (choosing 6 from 49) is $\dfrac{49!}{6!(49-6)!}$ = 13 983 816 to 1.

Conditional probability, P(A|B), indicates the probability of event A given that we know event B has occurred.

A Bayesian Network is a directed acyclic graph: A graph where the directions are links which indicate dependencies that exist between nodes. Nodes represent propositions about events or events themselves.

Conditional probabilities quantify the strength of dependencies.

Consider the following example:

The probability, P(E1) that my car won't start.

If my car won't start then it is likely that the battery is **flat** or the starting motor is broken. In order to decide whether to fix the car myself or send it to the garage I make the following decision:

- If the headlights do not work then the battery is likely to be flat so I fix it myself.
- If the starting motor is defective then send car to garage.
- If battery and starting motor are both **gone** send car to garage.

The network to represent this is as follows (Figure 3-2):

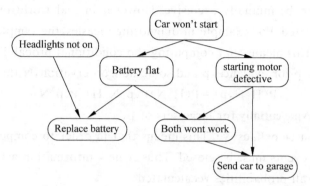

Figure 3-2　A simple Bayesian network

**Bayesian probabilistic inference**

Bayes' theorem can be used to calculate the probability that a certain event will occur or that a certain proposition is true. The theorem is stated as follows:

$P(B|A) = P(A|B)P(B)/P(A)$

$P(B)$ is called the prior probability of B. $P(B|A)$, as well as being called the conditional probability, is also known as the **posterior** probability of B.

$P(A \wedge B) = P(A|B)P(B)$

Note that due to the **commutativity** of $\wedge$, we can also write

$P(A \wedge B) = P(B|A)P(A)$

Hence, we can deduce: $P(B|A)P(A) = P(A|B)P(B)$

This can then be rearranged to give Bayes' theorem:

$P(B|A) = P(A|B)P(B)/P(A)$

Bayes Theorem states:

$$P(H_i|E) = \frac{P(E|H_i)P(H_i)}{\sum_{K=1}^{n} P(E|H_i)P(H_k)}$$

This reads that given some evidence E then probability that hypothesis $H_i$ is true is equal to the ratio of the probability that E will be true given $H_i$ times the **a priori** evidence on the probability of $H_i$ and the sum of the probability of E over the set of all hypotheses times the probability of these hypotheses.[1]

The set of all hypotheses must be mutually **exclusive** and **exhaustive**.

Thus to find if we examine medical evidence to diagnose an illness, we must know all the prior probabilities of finding symptom and also the probability of having an illness based on certain symptoms being observed.

Bayesian statistics lie at the heart of most statistical reasoning systems. How is Bayes theorem exploited?

The key is to formulate problem correctly:

P(A|B) states the probability of A given only B's evidence. If there is other relevant evidence then it must also be considered.

All events must be mutually exclusive. However in real world problems events are not generally unrelated. For example in diagnosing measles, the symptoms of spots and a fever are related. This means that computing the conditional probabilities gets complex.

In general if a prior evidence, p and some new observation, N then
$$P(H|N,p) = P(H|N)P(p|N,H)/P(p|N)$$
computing grows **exponentially** for large sets of p.

All events must be exhaustive. This means that in order to compute all probabilities the set of possible events must be closed. Thus if new information **arises** the set must be created **afresh** and all probabilities recalculated.

Thus simple Bayes rule-based systems are not suitable for uncertain reasoning.

- Knowledge acquisition is very hard.
- Too many probabilities needed—too large a storage space.
- Computation time is too large.
- Updating new information is difficult and time consuming.
- Exceptions like "one of the above" cannot be represented.
- Humans are not very good probability estimators.

However, Bayesian statistics still provide the core to reasoning in many uncertain reasoning systems with suitable enhancement to overcome the above problems.

Bayesian networks are also called Belief Networks or Probabilistic Inference Networks.

# Words

| | |
|---|---|
| probability[ˌprɔbəˈbiləti] n. 概率 | exclusive[ikˈskluːsiv] adj. 独有的,排外的,专一的 |
| ace[eis] n. 幺点 | |
| deck[dek] n. 一副扑克牌 | exhaustive[igˈzɔːstiv] adj. 彻底的,穷尽的 |
| factorial[fækˈtɔːriəl] n. 阶乘 | exponentially[ˌɛkspəˈnɛnʃ(ə)li] adv. 以指数方式 |
| flat[flæt] n. 少许电量 | |
| posterior[pɔˈstiəriə(r)] adj. 其次的,较后的 | arise[əˈraiz] v. 出现,上升 |
| commutativity[kəˌmjuːtəˈtiviti] n. 交换性 | afresh[əˈfreʃ] adv. 重新,再度 |

# Unit 3  Reasoning with Uncertainty

| | |
|---|---|
| spade suit | 黑桃花色 |
| a priori | 推理的 |

[1] **Original**: This reads that given some evidence E then probability that hypothesis $H_i$ is true is equal to the ratio of the probability that E will be true given $H_i$ times the a priori evidence on the probability of $H_i$ and the sum of the probability of E over the set of all hypotheses times the probability of these hypotheses.

**Translation**：这意味着给定一些证据 E，那么假设 $H_i$ 为真的概率等于 $p_1$ 与 $p_2$ 的比值，其中 $p_1$ 为给定 $H_i$ 时 E 为真的概率乘以 $H_i$ 的先验证据概率，$p_2$ 为所有假设集合中 E 的概率之和乘以这些假设的概率。

I. Read the following statements carefully, and decide whether they are true (T) or false (F) according to the text.

____1. Conditional probability, P(A|B), indicates the probability of event B given that we know event A has occurred.

____2. Probability theory is used to discuss events, categories, and hypotheses about which there is 100% certainty.

____3. Bayes' theorem can be used to calculate the probability that a certain event will occur or that a certain proposition is false.

____4. $C_m^n = \dfrac{m!}{n!(m-n)!}$

____5. We might write A→B, which means that if B is true, then A is true.

II. Choose the best answer to each of the following questions according to the text.

1. Which of the following is right? (　　)

   A. P(A|B) = P(B|A)P(A)/P(B)

   B. $C_m^n = \dfrac{m!}{n!(m-n)!}$

   C. P(U|V) = P(U∧V)/P(V)

   D. All of the above

2. What is Bayesian networks' other name? (　　)

   A. Telecommunication Networks

   B. Probabilistic Inference Networks

   C. Traffic Networks

   D. None of the above

3. Which of the followings are reasons for the simple Bayes rule-based systems that are not suitable for uncertain reasoning?（　　）

   A. Updating new information is difficult and time consuming.

   B. Too many probabilities needed—too large a storage space.

   C. Computation time is too large.

   D. All of the above

III. Fill in the blanks with the words or phrases chosen from the box. Change the forms where necessary.

| likely | despite | either | have | know |
| sure | otherwise | false | give | be |

**Understanding Truth-Values**

A truth-value is a label that is ___1___ to a statement (a proposition) that denotes the relation of the statement to truth.

In general, all statements, when worded properly, are ___2___ true or false (even if we don't know with certainty their truth-value, they are ultimately true or false ___3___ our ability to know for sure).

With that said, and as noted, humans can't ___4___ every truth for certain, and thus there are some "unknowns."

Thus, all truth-values can be transposed to a three-value form: True, ___5___, and Unknown (where unknown denotes our lack of knowledge, not a lack of truth).

Further, because there are things humans can't know for ___6___, we have to express the likelihood of some truths using multiple-value truth-values. For example: very ___7___ false, likely false, likely true, very likely true. Or, such is the case where the quality of a complex statement or argument is ___8___ considered, False, Mostly False, Half True, Mostly True, and True.

Additionally, because some statements have a variable truth-value that changes depending on context or on more information like "the cat is on the mat" (which is only true when the cat is on the mat, ___9___ it is the case that it is false) and x + 1 = 1 (which is only true when for example x = 0) we also ___10___ to consider "variable" or "conditional" truth-values (propositions that have truth-values that are not constant and instead depend on more information, but otherwise adhere to the other rules stated).

IV. Translate the following passage into Chinese.

**Probabilistic Reasoning**

Probabilistic reasoning is a way of knowledge representation where we apply the concept of probability to indicate the uncertainty in knowledge. In probabilistic reasoning, we combine probability theory with logic to handle the uncertainty.

We use probability in probabilistic reasoning because it provides a way to handle the uncertainty that is the result of someone's laziness and ignorance.

## Unit 3  Reasoning with Uncertainty

In the real world, there are lots of scenarios, where the certainty of something is not confirmed, such as "It will rain today," "behavior of someone for some situations," "A match between two teams or two players." These are probable sentences for which we can assume that it will happen but not sure about it, so here we use probabilistic reasoning.

## Part 2

# Simulated Writing: Communicating with Social Media

在科技交流方面，社交媒体可成为广泛的信息来源，其功能包括保持同事间的联系、从技术专家那里获取信息、了解顾客的喜好、与顾客分享最新的信息，以及在专业领域内进行广泛的联系。这里介绍几种常用的社交媒体（如博客、微博、社交网络站点、QQ 和微信）进行交流的方式。

### 1. 博客

一个博客（Blog）就是一个网页，它通常是由简短且经常更新的帖子构成；这些张贴的文章都按照年份和日期排列，也称为网络日志。博客的内容和目的有很大的不同，从对其他网站的超级链接和评论，有关公司、个人、构想的新闻到日记、照片、诗歌、散文，甚至科幻小说的发表或张贴都有。许多博客都是作者个人心中所想的发表，另一些博客则是一群人基于某个特定主题或共同利益领域的集体创作。

随着博客的快速扩张，它的目的与最初的浏览网页心得已相去甚远。网络上数以千计的博主发表和张贴博客的目的有很大的差异。不过，由于沟通方式比电子邮件、讨论群组更简单和容易，博客已成为家庭、公司、部门和团队之间越来越盛行的沟通工具，它也逐渐被应用在企业内部网络（Intranet）中。

有些公司利用博客在雇员间分享信息。这些仅在内部可访问的博客主要针对的是工程师、经理、行政人员和其他内部人员。公司所创建的内部博客可改进工作流程和提高士气。在大公司里，博客可作为电子邮件的一种替代方式，用于日常的内部交流。在网络上每个人都可上传消息，或对其他消息做出评论。在通过博客交流的环境里可以举行会议，不受面对面会议时间和场地的限制。还可以进行员工培训，传达公司发展的最新情况。博客对于协同工作特别有效。例如，公司软件部的人员可以创建论坛，讨论编程和测试的流程，然后部门人员可以加以衡量并提出建议。

公司所创建的外部博客可向客户提供对产品和服务的反馈渠道，改进营销和公共关系，提供及时的信息和更新，帮助大型组织的个性化。博客使公司有机会展示其人性和非正式的一面，对客户关注的问题做出亲切而快速的响应，让客户能提供想法和反馈。在公司的博客里语气十分关键，要表示友好、欢迎、真诚。博客应该邀请读者参与。

例如，迪士尼乐园（Disney Parks）创建了一个叫 Disney Parks Blog 的博客（如图 3-3 所示），尽管迪士尼的品牌本质上针对的是小孩子，但是这个博客的内容对那些渴望获取一些旅行小贴士的父母来说是很有用的。与大多数的博客不同的是，迪士尼没有强调其社交的形象，它只是具有 RSS（聚合内容）和电子邮件订阅的链接。迪士尼也没有将每一个帖子底部的 Twitter 和 Facebook 的社交共享按钮包括进来。有关这个博客的参与度似乎有点低但比较稳定，每个帖子都有几个评论，并且每个帖子都在任意位置最多有 100 个赞（like）。此外，通常缺失的元素是要号召导致读者想要评论的行动。还有一点就是，也许这个博客的受众就不是评论类的。

图 3-3　迪士尼乐园的博客

**2. 微博**

推特（Twitter）是美国的一个网站，它利用无线网络、有线网络的通信技术进行即时通信，是微博客（微博）的典型应用。它允许用户将自己的最新动态和想法以短信息的形式发送给手机和个性化的网站群，而不仅仅是发送给个人。

推特是一个可让你播报短消息给你的朋友或"跟随者（followers）"的在线服务，它也同样允许你指定你想跟随（follow）的推特用户，这样你就可以在一个页面上就能读取他们发布的信息。

推特提供以简洁形式进行实时发帖和更新的手段。个人以及公司、政府机构、其他组织通过推特简讯使朋友、商务关系人、顾客、公民获得信息并及时更新。推特被称为"小鸟叫声"（tweets），其写作方式就是简洁，每条消息限制在 140 个字符以内。

假如你在推特中输入一个项目，它们可以是私有的，只有当你的朋友获得你的允许才能查看；或者也可以是公开的，也就是说，所有知道你 Twitter ID 的人都可以读取或订阅你发布的消息。另外还有很重要的一点就是，推特是完全免费的。

推特可以用在很多地方，比如企业可以用它提供客户服务。图 3-4 就是 JetBlue 公司所发布的推特。

也可以在中国流行的新浪微博上进行撰写。

**3. 社交网络站点**

社交网络网站可将人们联系在一起，并将具有共同感兴趣的站点联系在一起。Facebook（脸谱）是最流行的社交网之一。Facebook 最初是为大学生而开发的，现在被朋友、家庭、专业协会、政治组织、非盈利机构、商业公司所使用。许多企业都有 Facebook 页面，用来强调或推销企业的一种特定产品或服务。例如招聘公司，可以创建 Facebook 页面

# Unit 3  Reasoning with Uncertainty

图 3-4  JetBlue 公司所发布的推特

进行招聘。非盈利机构可以利用 Facebook 页面分享信息、提供照片、允许其他 Facebook 用户进行评论,与朋友分享信息。NASA(美国航空航天局)的火星探测器"好奇"号有一个 Facebook 页面,曾经受到 50 万人的喜爱。

Facebook 最流行的使用是在个人之间,人们可以用它保持与亲友的联系。虽然用户可能认为自己的个人 Facebook 页面与工作没有什么关系,但应记住,雇主会经常查看 Facebook 和其他社交网站,作为他们招聘面试的一部分。

图 3-5 为一个典型的 Facebook 页面。

图 3-5  一个典型的 Facebook 页面

### 4. QQ

QQ 是一款基于 Internet 的即时通信(IM)软件。QQ 支持在线聊天、视频通话、点对点断点续传文件、共享文件、网络硬盘、自定义面板、QQ邮箱等多种功能，并可与多种通信终端相连。目前QQ已经覆盖Microsoft Windows、macOS、Android、iOS等多种主流平台。QQ不仅仅是个人交流的平台，也是现代企业常用的推广工具。

图3-6是一个使用QQ进行交流的例子。

图3-6 一个使用QQ进行交流的例子

### 5. 微信

微信（wechat）是一个为智能终端提供即时通信服务的免费应用程序。微信支持跨通信运营商、跨操作系统平台，通过网络快速发送免费的（需消耗少量网络流量）语音短信、视频、图片和文字；微信提供公众平台、朋友圈、消息推送等功能，用户可以通过"摇一摇""搜索号码"、扫描二维码等方式添加好友和关注公众平台，也可将内容分享给好友或朋友圈。

其实微信相当于另一个QQ。但与QQ不同的是，它在交友方面的表现更具时效性，也更强大。不论是安卓系统还是苹果系统的手机用户，只要安装了微信，就可以进行跨手机平台的畅通聊天。

图3-7是一个使用微信进行交流的例子。

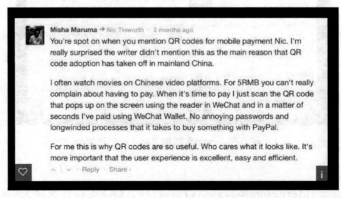

图3-7 一个使用微信进行交流的例子

# Unit 3  Reasoning with Uncertainty

## Part 3

# Listening & Speaking

### Dialogue: Reasoning with Uncertainty

(Today is the first day after the National Day holiday. Henry met Mark and Sophie in the hall.)

在线音频

Henry: Hi, Mark and Sophie. How was your National Day holiday?

Mark: It was not bad. During this holiday, I studied reasoning with uncertainty in AI. It's very interesting.

Henry: Reasoning with uncertainty? Don't we often reason with certainty?

Mark: Yes, you are right. Regarding knowledge representation using first-order logic and propositional logic with certainty, we are sure about the predicates. With this knowledge representation, we might write A→B, which means if A is true then B is true, but consider a situation where we are not sure about whether A is true or not then we cannot express this statement, this situation is called uncertainty.

Henry: So how to represent uncertain knowledge?

Sophie: Well, to represent uncertain knowledge, where we are not sure about the predicates, we need uncertain reasoning or probabilistic reasoning.

Henry: Probabilistic reasoning?

Sophie: Yes. Probabilistic reasoning is a way of knowledge representation where we apply the concept of probability to indicate the uncertainty in knowledge. In probabilistic reasoning, we combine probability theory with logic to handle the uncertainty.

Mark: Furthermore,[1] we use probability in probabilistic reasoning because it provides a way to handle the uncertainty that is the result of someone's laziness and ignorance.

Sophie: Actually, in the real world, there are lots of scenarios, where the certainty of something is not confirmed, such as "It will rain today," "behavior of someone for some situations," "A match between two teams or two players." These are probable sentences for which we can

[1] Replace with:
1. In addition,
2. Moreover,
3. What is more,
4. Additionally,
5. Besides,
6. Plus,
7. Also,

assume that it will happen but not sure about it, so here we use probabilistic reasoning.

**Henry:** So are there any ways to solve problems with uncertain knowledge?

**Sophie:** Of course. In probabilistic reasoning, there are two ways to solve problems with uncertain knowledge: Bayes' rule and Bayesian Statistics.

**Mark:** And as probabilistic reasoning uses probability and related terms, first, let's talk about probability. Probability can be defined as a chance that an uncertain event will occur. It is the numerical measure of the likelihood that an event will occur. The value of probability always remains between 0 and 1 that represent ideal uncertainties.

**Sophie:** That's right. Conditional probability is the probability of one event occurring with some relationship to one or more other events.

**Henry:** I'm sorry, I'm afraid I have to learn probability and other knowledge of reasoning with uncertainty a little bit in advance and then discuss it with you later, and otherwise [2] it is very difficult for me to follow you. Thanks anyway.

**Mark:** Looking forward to seeing you again!

**Sophie:** Good luck to you!

[2] Replace with:
1. or else
2. or
3. if not
4. or then

##  Exercises

Work in a group, and make up a similar conversation by replacing the statements with other expressions on the right side.

## Listening Comprehension: Fuzzy Logic

*Listen to the article and answer the following 3 questions based on it. After you hear a question, there will be a break of 15 seconds. During the break, you will decide which one is the best answer among the four choices marked (A), (B), (C) and (D).*

**Questions**

1. When was the idea of fuzzy logic first advanced by Dr. Lotfi Zadeh of the University of California at Berkeley? (    )

   (A) In 1940s
   (B) In 1950s
   (C) In 1960s

(D) In 1970s

2. Which of the following is right?（　　）

(A) Fuzzy logic seems closer to the way our brains work.
(B) Fuzzy logic is essential to the development of human-like capabilities for AI.
(C) Natural language is not easily translated into the absolute terms of 0 and 1.
(D) All of the above

3. Which of the following is not right?（　　）

(A) Fuzzy logic includes 0 and 1 as extreme cases of truth.
(B) Fuzzy logic is not an approach to computing based on the usual "true or false" (1 or 0) Boolean logic on which the modern computer is based.
(C) Fuzzy logic is an approach to computing based on "degrees of truth".
(D) None of the above

## Words

advance[əd'vɑːns] v. 提出
aggregate['æɡrigət] v. 合计
threshold['θreʃhəuld] n. 阈值，临界值

## Phrases

in between　在中间
and so　因此，所以
motor reaction　动作反应，运动反应

## Dictation：Bayesian Network

在线音频

*This article will be played three times. Listen carefully, and fill in the numbered spaces with the appropriate words you have heard.*

A Bayesian network, Bayes network, belief network, decision network, Bayes(ian) model or probabilistic directed acyclic ___1___ model is a probabilistic graphical model (a type of statistical model) that represents a set of variables and their conditional dependencies via a Directed Acyclic Graph (DAG). For example, a Bayesian network could ___2___ the probabilistic relationships between diseases and symptoms. ___3___ symptoms, the network can be used to compute the probabilities of the presence of various diseases.

Efficient ___4___ can perform inference and learning in Bayesian networks. Bayesian networks that model ___5___ of variables (e. g. speech signals or **protein** sequences) are called dynamic Bayesian networks. Generalizations of Bayesian networks that can represent and solve decision problems under uncertainty are ___6___ influence diagrams.

The Bayesian methods have a number of advantages that ___7___ their suitability in uncertainty management. One of the advantages is that most ___8___ is their **sound** theoretical foundation in probability theory. Thus, they are ___9___ the most mature of all of the uncertainty reasoning methods.

While Bayesian methods are more developed than the other uncertainty methods, they are not without ___10___ .

1. They require a significant amount of probability data to ___11___ a knowledge base. Furthermore, human experts are normally uncertain and ___12___ about the probabilities they are providing.

2. What are the relevant prior and conditional probabilities ___13___ on? If they are statistically based, the sample sizes must be ___14___ so the probabilities obtained are accurate. If human experts have provided the values, are the values consistent and comprehensive?

3. Often the type of relationship between the ___15___ and evidence is important in determining how the uncertainty will be ___16___ . Reducing these associations to simple numbers removes relevant information that might be needed for successful reasoning about the uncertainties. For example, Bayesian-based medical diagnostic systems have ___17___ to gain acceptance because physicians ___18___ systems that cannot provide explanations describing how a conclusion was reached (a feature difficult to provide in a Bayesian-based system).

4. The reduction of the associations to numbers also ___19___ using this knowledge within other tasks. For example, the associations that would enable the system to explain its reasoning to a user are lost, as is the ability to browse through the ___20___ of evidences to hypotheses.

## ▶▶▶ Words

| protein['prəuti:n] n. 蛋白质 | sound[saund] adj. 健全的,(非正式)非常棒的 |

# Unit 4

# Search Methods in Artificial Intelligence

# Part 1

# Reading & Translating

## Section A: Heuristic Search

Heuristics are problem-solving strategies which in many cases find a solution faster than **uninformed search**. However, this is not guaranteed. Heuristic search could require a lot more time and can even result in the solution not being found.

We humans successfully use heuristic processes for all kinds of things. When buying vegetables at the supermarket, for example, we judge the various options for a pound of **strawberries** using only a few simple criteria like price, appearance, source of production, and **trust in** the seller, and then we **decide on** the best option by **gut** feeling. It might theoretically be better to **subject** the strawberries **to** a basic chemical analysis before deciding whether to buy them. For example, the strawberries might be poisoned. If that were the case the analysis would have been worth the trouble. However, we do not carry out this kind of analysis because there is a very high probability that our heuristic selection will succeed and will quickly get us to our goal of eating **tasty** strawberries.

Heuristic decisions are closely linked with the need to make real-time decisions with limited resources. In practice a good solution found quickly is **preferred over** a solution that is optimal, but very expensive to derive.

A heuristic evaluation **function** $f(s)$ for states is used to mathematically model a heuristic. The goal is to find, with little effort, a solution to the stated search problem with minimal total cost. Please note that there is a subtle difference between the effort to find a solution and the total cost of this solution. For example it may take Google Maps half a second's worth of effort to find a route from the City Hall in San Francisco to Tuolumne Meadows in Yosemite National Park, but the ride from San Francisco to Tuolumne Meadows by car may take four hours and some money for gasoline etc. (total cost).

Next we will modify the breadth-first search algorithm by adding the evaluation function to it. The currently open nodes are no longer expanded left to right by row, but rather according to their heuristic **rating**. From the set of open nodes, the node with the minimal rating is always expanded first. This is achieved by immediately evaluating nodes as they are expanded and sorting them into the list of open nodes. The list may then contain nodes from different depths in the tree.

Because heuristic evaluation of states is very important for the search, we will differentiate from now on between states and their associated nodes. The node contains the state and further information relevant to the search, such as its depth in the search tree and the heuristic rating of the state. As a result, the function "Successors", which generates the successors (children) of a node, must also immediately calculate for these successor nodes their heuristic ratings as a component of each node. We define the general search algorithm HEURISTICSEARCH in Figure 4-1.

```
HEURISTICSEARCH(Start, Goal)
NodeList=[Start]
While true
    If NodeList=∅ Return("No solution")
    Node= First(NodeList)
    NodeList= Rest(Node List)
    If GoalReached(Node, Goal) Return(Solution found", Node)
    Nodelist= SortIn( Successors(Node), NodeList)
```

Figure 4-1 The general search algorithm HEURISTICSEARCH

The node list is initialized with the starting nodes. Then, in the loop, the first node from the list is removed and tested for whether it is a solution node. If not, it will be expanded with the function "Successors" and its successors added to the list with the function "SortIn". "SortIn(X,Y)" inserts the elements from the unsorted list X into the ascendingly sorted list Y. The heuristic rating is used as the sorting key.

Thus it is guaranteed that the best node (that is, the one with the lowest heuristic value) is always at the beginning of the list. When sorting in a new node from the node list, it may be advantageous to check whether the node is already available and, if so, to delete the duplicate.

Depth-first and breadth-first search also happen to be special cases of the function HEURISTICSEARCH. We can easily generate them by plugging in the appropriate evaluation function.

The best heuristic would be a function that calculates the actual costs from each node to the goal. To do that, however, would require a **traversal** of the entire search space, which is exactly what the heuristic is supposed to prevent. Therefore we need a heuristic that is fast and simple to compute. How do we find such a heuristic?

An interesting idea for finding a heuristic is simplification of the problem. The original task is simplified enough that it can be solved with little computational cost. The costs from a state to the goal in the simplified problem then serve as an estimate for the actual problem. This cost estimate function we denote h.

  **Words**

| | |
|---|---|
| strawberry[ˈstrɔːbəri] n. 草莓 | function[ˈfʌŋkʃn] n. 函数 |
| gut[gʌt] adj. 本能的, 直觉的, 简单的 | rating[ˈreitiŋ] n. 等级评定, 等级 |
| tasty[ˈteisti] adj. 美味的 | traversal[trəˈvɜːs(ə)l] n. 遍历 |

## Phrases

uninformed search　盲目搜索
trust in　信任
decide on　选定,决定
subject to　受制于
prefer over　更喜欢

## Exercises

I. Read the following statements carefully, and decide whether they are true（T）or false（F）according to the text.

　　____ 1. It is guaranteed that the best node (that is, the one with the lowest heuristic value) is always at the end of the list.

　　____ 2. There is a delicate difference between the effort to find a solution and the total cost of this solution.

　　____ 3. The worst heuristic would be a function that calculates the actual costs from each node to the goal.

　　____ 4. Finding a good solution quickly is better than finding an optimal solution.

　　____ 5. Modeling a heuristic mathematically can use a heuristic evaluation function f(s) for states.

II. Choose the best answer to each of the following questions according to the text.

1. Which of the following is very expensive to derive?（　　）

　　A. A solution that is optimal

　　B. A good solution found quickly

　　C. A good solution found slowly

　　D. All of the above

2. Which of the following is guaranteed?（　　）

　　A. The best node (that is, the one with the lowest heuristic value) is always in the middle of the list.

　　B. The best node (that is, the one with the lowest heuristic value) is always at the end of the list.

　　C. The best node (that is, the one with the lowest heuristic value) is always at the beginning of the list.

　　D. None of the above

3. What kind of difference is there between the effort to find a solution and the total cost of this solution according to this text?（　　）

　　A. A subtle difference

B. A simple difference
C. An obvious difference
D. None of the above

III. Fill in the numbered spaces with the words or phrases chosen from the box. Change the forms where necessary.

> appear  outperform  base  desire  other  performance
> technique  approach  commonsense  evolved

**Behavior-Based Intelligence**

Early work in artificial intelligence __1__ the subject in the context of explicitly writing programs to simulate intelligence. However, many argue today that human intelligence is not __2__ on the execution of complex programs but instead by simple stimulus-response functions that have __3__ over generations. This theory of "intelligence" is known as behavior-based intelligence because "intelligent" stimulus response functions __4__ to be the result of behaviors that caused certain individuals to survive and reproduce while __5__ did not.

Behavior-based intelligence seems to answer several questions in the artificial intelligence community such as why machines based on the von Neumann architecture easily __6__ humans in computational skills but struggle to exhibit __7__. Thus behavior-based intelligence promises to be a major influence in artificial intelligence research. Behavior-based __8__ have been applied in the field of artificial neural networks to teach neurons to behave in __9__ ways, in the field of genetic algorithms to provide an alternative to the more traditional programming process, and in robotics to improve the __10__ of machines through reactive strategies.

IV. Translate the following passage into Chinese.

**Travelling Salesman Problem**

A salesman wants to travel to N cities (he should pass by each city). How can we order the cities so that the salesman's journey will be the shortest? The objective function to minimize here is the length of the journey (the sum of the distances between all the cities in a specified order).

To start solving this problem, we need:

- Configuration setting: This is the permutation of the cities from 1 to N, given in all orders. Selecting an optimal one between these permutations is our aim.
- Rearrangement strategy: The strategy that we will follow here is replacing sections of the path, and replacing them with random ones to retest if this modified one is optimal or not.
- The objective function (which is the aim of the minimization): This is the sum of the distances between all the cities for a specific order.

## Section B: Genetic Algorithms

The underlying idea in all evolutionary algorithms is that we take a **population** of individuals and apply the natural selection process. We start with a set of randomly selected individuals and then identify the strongest among them. The strength of each individual is determined using a **fitness** function that's predefined. In a way, we use the survival of the fittest approach.

We then take these selected individuals and create the next generation of individuals by **recombination** and **mutation**. For now, let's think of these techniques as mechanisms to create the next generation by treating the selected individuals as parents.

Once we execute recombination and mutation, we create a new set of individuals who will compete with the old ones for a place in the next generation. By discarding the weakest individuals and replacing them with **offspring**, we are increasing the overall fitness level of the population. We continue to iterate until the desired overall fitness is achieved.

A genetic algorithm [1] is an evolutionary algorithm where we use a heuristic to find a string of bits that solves a problem (Figure 4-2). We continuously iterate on a population to arrive at a solution. We do this by generating new populations containing stronger individuals. We apply probabilistic operators such as selection, **crossover**, and mutation in order to generate the next generation of individuals. The individuals are basically strings, where every string is the encoded version of a potential solution.

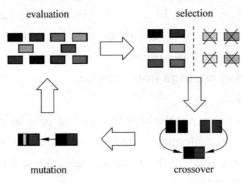

Figure 4-2　A genetic algorithm

A fitness function is used that evaluates the fitness measure of each string telling us how well suited it is to solve this problem. This fitness function is also referred to as an evaluation function. Genetic algorithms apply operators that are inspired from nature, which is why the **nomenclature** is closely related to the terms found in biology.

In order to build a genetic algorithm, we need to understand several key concepts and terminology. These concepts are used extensively throughout the field of genetic algorithm to build solutions to various problems. One of the most important aspects of genetic algorithms is the randomness. In order to iterate, it relies on the random sampling of individuals. This means that the process is non-deterministic. So, if you run the same algorithm multiple times, you might **end up with** different solutions.

Let's talk about population. A population is a set of individuals that are possible candidate solutions. In a genetic algorithm, we do not maintain a single best solution at any given stage. It maintains a set of potential solutions, one of which is the best. But the other solutions play an important role during the search. Since we have a population of solutions, it is less likely that will get **stuck** in a **local optimum**. Getting stuck in the local optimum is a classic problem faced by other optimization techniques.

Now that we know about population and the **stochastic** nature of genetic algorithms, let's talk about the operators. In order to create the next generation of individuals, we need to make sure that they come from the strongest individuals in the current generation. Mutation is one of the ways to do it. A genetic algorithm makes random changes to one or more individuals of the current generation to yield a new candidate solution. This change is called mutation. Now this change might make that individual better or worse than existing individuals.

The next concept here is recombination, which is also called crossover. This is directly related to the role of reproduction in the evolution process. A genetic algorithm tries to combine individuals from the current generation to create a new solution. It combines some of the features of each parent individual to create this offspring. This process is called crossover. The goal is to replace the weaker individuals in the current generation with offspring generated from stronger individuals in the population.

In order to apply crossover and mutation, we need to have selection criteria. The concept of selection is inspired by the theory of natural selection. During each iteration, the genetic algorithm performs a selection process. The strongest individuals are chosen using this election process and the weaker individuals are terminated. This is where the survival of the fittest concept **comes into play**. The selection process is carried out using a fitness function that computes the strength of each individual.

## Words

population[ˌpɒpjuˈleɪʃn] *n.* 种群
fitness[ˈfɪtnəs] *n.* 适应度，健康
recombination [ˌriːkɒmbɪˈneɪʃ(ə)n; ˌriːkɒmb-] *n.* 重组
mutation[mjuːˈteɪʃn] *n.* 变异，突变
offspring[ˈɒfsprɪŋ] *n.* 后代，子孙

crossover[ˈkrɒsəʊvə(r)] *n.* 组合交叉
nomenclature[nəˈmeŋklətʃə(r)] *n.* 术语，命名法
stuck[stʌk] *adj.* （因困难）无法继续的，停滞不前的
stochastic[stəˈkæstɪk] *n.* 随机的，猜测的

 **Phrases**

end up with 以……告终，以……结束
local optimum 局部最优，局部优化
come into play （使）开始起作用

 **Notes**

［1］ 遗传算法（genetic algorithm）是计算机科学人工智能领域中用于解决最优化问题的一种搜索启发式算法，是进化算法的一种。这种启发式算法通常用来生成有用的解决方案来优化和搜索问题。进化算法最初是借鉴了进化生物学中的一些现象而发展起来的，这些现象包括遗传、突变、自然选择以及杂交等。遗传算法在适应度函数选择不当的情况下有可能收敛于局部最优，而不能达到全局最优。

 **Exercises**

I. Read the following statements carefully, and decide whether they are true（T）or false（F）according to the text.

　　____ 1. The concept of selection is inspired by the theory of artificial selection.
　　____ 2. Crossover is also called recombination.
　　____ 3. Mutation might make that individual better or worse than existing individuals.
　　____ 4. An evaluation function is also referred to as a fitness function.
　　____ 5. A genetic algorithm is a kind of heuristic algorithm.

II. Choose the best answer to each of the following questions according to the text.

1. Which of the following probabilistic operators do we apply in order to generate the next generation of individuals?（　　）
   A. Selection
   B. Crossover
   C. Mutation
   D. All of the above

2. Which kind of algorithm does a genetic algorithm belong to?（　　）
   A. K-means
   B. Heuristic
   C. DQN
   D. Q-learning

3. What do we need to have in order to apply crossover and mutation?（　　）
   A. Selection criteria
   B. Recombination
   C. Random changes
   D. None of the above

## Unit 4  Search Methods in Artificial Intelligence

III. Fill in the numbered spaces with the words or phrases chosen from the box. Change the forms where necessary.

> choose  whether  generate  call  turn
> wide  iterative  use  available  long

### Hill Climbing

Hill climbing is an optimization technique that is ___1___ to find a "local optimum" solution to a computational problem. It starts off with a solution that is very poor compared to the optimal solution and then iteratively improves from there. It does this by ___2___ "neighbor" solutions which are relatively a step better than the current solution, picks the best and then repeats the process until it arrives at the most optimal solution because it can no ___3___ find any improvements.

Variants:
- Simple — The first closest node or solution to be found is ___4___.
- Steepest ascent — All ___5___ successor solutions are considered and then the closest one is selected.
- Stochastic — A neighbor solution is selected at random, and it is then decided ___6___ or not to move on to that solution based on the amount of improvement over the current node.

Hill climbing is done ___7___ — it goes through an entire procedure and the final solution is stored. If a different iteration finds a better final solution, the stored solution or state is replaced. This is also ___8___ shotgun hill climbing, as it simply tries out different paths until it hits the best one, just like how a shotgun is inaccurate but may still hit its target because of the ___9___ spread of projectiles. This works very well in many cases because at it ___10___ out, it is better to spend CPU resources exploring different paths than carefully optimizing from an initial condition.

IV. Translate the following passage into Chinese.

### Evolutionary Programming

When applied to the task of program development, the genetic algorithm approach is known as evolutionary programming. Here the goal is to develop programs by allowing them to evolve rather than by explicitly writing them. Researchers have applied evolutionary programming techniques to the program development process using functional programming languages. The approach has been to start with a collection of programs that contain a rich variety of functions. The functions in this starting collection form the "gene pool" from which future generations of programs will be constructed. One then allows the evolutionary process to run for many generations, hoping that by producing each generation from the best performers in the previous generation, a solution to the target problem will evolve.

# Part 2

# Simulated Writing: Using Presentation Software to Write

使用 Microsoft PowerPoint（最常用）或其他软件，如 Apple Keynote 或 OpenOffice Impress，能很快、很容易地创建形象化元素，并且可将图像、声音、电影、动画、视频的网上链接包括进来。一种新的软件 Prezi 可以提供更多的选择。

在充满图像、视频的电子通信世界里，PowerPoint 或类似的软件常被认为是企业不可或缺的工具。

使用演示软件可以做以下事情：
- 用不同的颜色、底纹、纹理创建幻灯片。
- 创建绘画或曲线图，导入剪贴图、照片、其他图像。
- 创建有动画效果的文本和图像。例如，每点击一次进入一个符号列表项，或者图形中的纵向条形和横向条形逐个加亮，强调数据中的某些特征。
- 创建幻灯片之间的动态过渡，例如，一张幻灯片向右侧退出，另一张从左侧进入。
- 将每张幻灯片放大。
- 将幻灯片按不同顺序排列。
- 精确设定演讲时间。
- 在计算机屏幕上、大型投影仪上、网页上、投影机上显示出演示，或打印出来作为讲义分发。

下面以 PowerPoint 为例，具体介绍如何制作幻灯片。

1. 简介

PowerPoint 演示文稿（PPT）和海报演示很类似，唯一不同的是 PPT 是在计算机上用幻灯片来演示，而不是用真正的海报。这种形式往往用来配合口头报告，可以使报告更具说服力。还可以加入声音和视觉媒体。幻灯片演示通常用于与一个大的群体分享信息，如专业会议、课堂演示和会议等。幻灯片演示应该更像是报告的提纲。在幻灯片演示中有 3 个主要元素：文本、图片、表格。文本用于突出主要观点以及关键术语和概念。

2. 一些准备工作

在开始做 PPT 之前，应该先考虑清楚演示的几个关键部分：受众、目的（说服性、增长知识性）、主题以及陈述。因为一个好的 PPT 设计应该适合用于演示的场合，因此在开始制作 PPT 之前，尽量搞清使用何种措辞有助于成功演示。

3. 主要组成部分

PPT 的幻灯片应该更像一个提纲。文字通常都是将主要内容列出来，而不是完整的句子。以下是几个可以在幻灯片中展示的内容：
- 图和/或表格；
- 解释；
- 列表；
- 基本事实；
- 必要的图片。

4. 内容安排

PPT 内容的顺序应该取决于受众所需。无论做什么，都要谨慎地组织 PPT 并井井有条地陈述论点，这样受众才会信服我们的论点。以下是一些可以参照的格式。
- 概述、主体、结论（此类格式通常用于帮助受众弄清演示内容，否则受众可能会被复杂的论点搞糊涂）；
- 轶事、内容、结论（此类格式主要为了防止受众在开始演示之前就已经厌烦）；
- 计划、好处、轶事（此类格式主要用于介绍一些新鲜事物，而且所面对的受众要求我们的演示简短且切题）。

5. 其他有用的信息

提倡的内容：
- 为整个 PPT 选择一个统一背景；
- 使用简单清晰的字体；
- 保证字体足够大，使后排的受众也能看到；
- 在列表中使用项目符号和一致的短语结构；
- 列出基本信息即可，运用关键词来引导读者或受众贯穿整个演示过程；
- 使用直接、简洁的语言，将文字数量减到最少；
- 必要时提供相关的解释；
- 使用空格来隔开文本或图片表格之类的可视组件；
- 保证每张幻灯片之间的联系符合逻辑；
- 为每张幻灯片加上标题。

避免的内容：
- 在幻灯片中堆满了图表；
- 使用复杂的字体；
- 添加不相干的信息；
- 写下所有要说的话；
- 使用分散受众注意力的图片；
- 颜色搭配难以阅读，如在蓝色背景上使用黑色的字体，尽量使用对比度高的组合。

使用 PowerPoint 展示的一个例子如下：

# Introduction to Extreme Programming (XP)

Source: ExtremeProgramming.org home

## What is XP

- Extreme Programming (XP) is actually a deliberate and disciplined approach to software development. It was based on observations of what made computer programming faster and what made it slower. About eight years old, it has already been proven at many companies of all different sizes and industries world wide.
- XP is an important new methodology for two reasons:
  - it is a re-examination of software development practices that have become standard operating procedures.
  - It is one of several new lightweight software methodologies created to reduce the cost of software.
- XP is successful because it emphasizes:
  - Customer involvement and satisfaction
  - Team work
- XP improves a software project in four essential ways:
  - Communication
  - Simplicity
  - Feedback
  - Courage

# Unit 4  Search Methods in Artificial Intelligence

## When to use XP

- Dynamic requirements
- High project risks
- Small groups of programmers
- Testability
- Productivity

## The Rules and Practices of XP

- Planning
    - User stories are written.
    - Release planning creates the schedule.
    - *Make frequent small releases.*
    - The Project Velocity is measured.
    - The project is divided into iterations.
    - Iteration planning starts each iteration.
    - Move people around.
    - A stand-up meeting starts each day.
    - Fix XP when it breaks.
- Designing
    - Simplicity.
    - Choose a system metaphor.
    - Use CRC cards for design sessions.
    - Create spike solutions to reduce risk.
    - No functionality is added early.
    - Refactor whenever and wherever possible.

## The Rules and Practices of XP *(cont.)*

- Coding
    - The customer is always available.
    - Code must be written to agreed standards.
    - Code the unit test first.
    - All production code is pair programmed.
    - Only one pair integrates code at a time.
    - Integrate often.
    - Use collective code ownership.
    - Leave optimization till last.
    - No overtime.
- Testing
    - All code must have unit tests.
    - All code must pass all unit tests before it can be released.
    - When a bug is found tests are created.
    - Acceptance tests are run often and the score is published.

## XP Map

- The XP Map shows how they work together to form a development methodology. Unproductive activities have been trimmed to reduce costs and frustration.

Figure 1 XP Map

## What We Have Learned About XP

- Release Planning
- Simplicity
- System Metaphor
- Pair Programming
- Integrate Often
- Optimize Last
- Unit Tests
- Acceptance Tests

## More Information

- Web Sites
    - The Portland Pattern Repository
    - XProgramming.com
    - XP Developer
    - ...
- Books
    - *Extreme Programming Explained: Embrace Change.* By Kent Beck
    - *Refactoring Improving the Design of Existing Code.* By Martin Fowler
    - *Extreme Programming Installed.* By Ron Jeffries, Chet Hendrickson, and Ann Anderson
    - ...

# Part 3

# Listening & Speaking

在线音频

## Dialogue: Search Methods in Artificial Intelligence

(Henry would like to know more about Search Methods in Artificial Intelligence. Then he asks for Mark and Sophie about it after class.)

**Henry:** Exactly what are search methods in Artificial Intelligence, Mark and Sophie?

**Sophie:** Well, search is inherent to the problems and methods of Artificial Intelligence (AI). That is because AI problems are intrinsically complex. Efforts to solve problems with computers which humans can routinely solve by employing **innate** cognitive abilities, pattern recognition, perception and experience, **invariably** must turn to considerations of search.

**Mark:** And all search methods essentially fall into one of two categories: (a) Exhaustive (blind) or uninformed methods and (b) Heuristic or informed methods.

**Henry:** Could you please [1] talk about uninformed methods in detail?

**Mark:** Sure. The Depth First Search (DFS) is one of the most basic and fundamental Blind Search Algorithms. It is for those who want to probe deeply down a potential solution path in the hope that solutions do not lie too deeply down the tree.

**Sophie:** And Breadth First Search always explores nodes closest to the root node first, thereby visiting all nodes of a given length first before moving to any longer paths. It pushes uniformly into the search tree. Breadth first search is most effective when all paths to a goal node **are of** uniform depth.

[1] Replace with:
1. Would you kindly...
2. Would you please...
3. Could you kindly...

# Unit 4  Search Methods in Artificial Intelligence

**Henry:** So how about [2] heuristic or informed methods?

**Sophie:** Heuristic search refers to a search strategy that attempts to optimize a problem by iteratively improving the solution based on a given heuristic function or a cost measure. A heuristic search method does not always guarantee to find an optimal or the best solution, but may instead find a good or acceptable solution within a reasonable amount of time and memory space.

**Mark:** Besides, several commonly used heuristic search methods include hill climbing methods, Greedy search (best first search), the A* algorithm, **simulated-annealing**, and genetic algorithms. A classic example of applying heuristic search is the traveling salesman problem.

**Henry:** Quite interesting. So what are best first search and the A* algorithm?

**Sophie:** Best first search expands the node that appears to be closest to goal and A* search minimizes the total estimated solution cost that includes cost of reaching a state and cost of reaching goal from that state.

**Henry:** Could we solve all the problems using these algorithms?

**Mark:** Depending on the problem, an Artificial Intelligence can use many other algorithms involving Machine Learning, Bayesian networks, Markov models, etc.

**Sophie:** Furthermore, selecting the right search strategy for your Artificial Intelligence can greatly amplify the quality of results. This involves formulating the problem that your AI is going to solve in the right way.

**Henry:** Honestly, it is really nice to talk with you about this kind of topic. Perhaps next time. Many thanks to you.

**Mark & Sophie:** No problem.

[2] Replace with:
1. what about
2. what do you think of

## Exercises

Work in a group, and make up a similar conversation by replacing the statements with other expressions on the right side.

## Words

| | |
|---|---|
| innate[iˈneit] *adj.* 先天的,固有的,与生俱来的<br>invariably[inˈveəriəbli] *adv.* 总是,不变 | 地,一定地<br>simulated-annealing 模拟退火法 |

## Phrases

be of    具有……性质

## Listening Comprehension: A* Search

*Listen to the article and answer the following 3 questions based on it. After you hear a question, there will be a break of 15 seconds. During the break, you will decide which one is the best answer among the four choices marked (A), (B), (C) and (D).*

**Questions**

1. A* algorithm is widely used in which areas?(        )

    (A) Graph-searching

    (B) Pathfinding

    (C) Graph traversal

    (D) All of the above

2. Which of the following is not right?(        )

    (A) A* is an extension of Dijkstra's algorithm with some characteristics of deapth-first search (BFS).

    (B) A* is an extension of Dijkstra's algorithm with some characteristics of breadth-first search (DFS).

    (C) A* works by making a lowest-cost path tree from the start node to the target node.

    (D) None of the above

3. Which of the following is right?(        )

    (A) A* is a heuristic function.

    (B) A* is not necessarily provably correct.

# Unit 4　Search Methods in Artificial Intelligence

　　(C) A* is different from an algorithm in that a heuristic is more of an estimate.
　　(D) All of the above

 **Words**

| pathfinding[ˈpæθˌfaindiŋ] n. 寻找目标，探险 |

 **Phrases**

| more of　更大程度上 |

## Dictation: Heuristic Search Techniques

*This article will be played three times. Listen carefully, and fill in the numbered spaces with the appropriate words you have heard.*

　　More often, there are so many possible options to solve a given problem that no ___1___ can be developed to find a right solution. Also, going through every single solution is not possible because it is **prohibitively** ___2___. In such cases, we rely on a **rule of thumb** that helps us **narrow down** the search by eliminating the options that are ___3___ wrong. This rule of thumb is called a heuristic. The method of using heuristics to guide our search is called heuristic search.

　　Heuristic techniques are very **handy** because they help us ___4___ up the process. Even if the heuristic is not able to completely eliminate some ___5___, it will help us to **order** those options so that we are more likely to get to the better solutions first.

　　If you are familiar with computer science, you should have heard about search techniques like Depth First Search (DFS), Breadth First Search (BFS), and Uniform Cost Search (UCS). These are search techniques that are ___6___ used on graphs to get to the solution. These are examples of uninformed search. They do not use any prior information or rules to eliminate some paths. They check all the plausible paths and pick the ___7___ one.

　　Heuristic search, on the other hand, is called **informed search** because it uses ___8___ information or rules to eliminate unnecessary paths. Uninformed search techniques do not take the goal into ___9___. These techniques don't really know where they are trying to go unless they just **stumble upon** the ___10___ in the process.

　　In the graph problem, we can use heuristics to guide the search. For example, at each ___11___, we can define a heuristic function that returns a score that represents the ___12___ of the cost of the path from the current node to the goal. By defining this heuristic function, we are informing the search technique about the right ___13___ to reach the goal. This will allow the algorithm to identify which neighbor will ___14___ to the goal.

We need to ___15___ that heuristic search might not always find the most optimal solution. This is because we are not ___16___ every single possibility and we are relying on a heuristic. But it is guaranteed to find a good solution in a ___17___ time, which is what we expect from a practical solution. In real-world ___18___, we need solutions that are fast and effective. Heuristic searches provide an ___19___ solution by arriving at a reasonable solution quickly. They are used in cases where the problems cannot be solved in any other way or would take a really long time to ___20___.

## Words

| prohibitively [prə'hibətivli] *adv.* 过高地，过分地 | handy ['hændi] *adj.* 有用的，便利的 order ['ɔːdə(r)] *v.* 整理 |
|---|---|

## Phrases

rule of thumb  经验法则
narrow down  缩小
informed search  启发式搜索
stumble upon  偶然发现

# Unit 5

## Machine Learning

# Part 1

# Reading & Translating

## Section A: Decision Tree in Machine Learning

Decision Tree in Machine Learning is used for supervised learning [classification and **regression**]. Decision Tree exploits correlation between features and non-linearity in the features.

Wondering what a Decision Tree would be? You might have come across the **programmatic** representation of a decision tree which is a nested if-else.

Let us consider the following pseudo logic, where we are trying to classify the given living-thing into either human, bird or plant:

```
if(displacement is present){
   if(wings are present AND feathers are present){
      living-thing is bird
   } else if(hands are present){
      living-thing is human
   }
} else if(displacement is absent){
   living-thing is plant
}
```

In the above pseudo code, output variable is category of living-thing whose value could be human or bird or plant. Input variable is living-thing. Features of input data taken into consideration are **displacement** [whose values are present/absent], wings [whose values are present/absent], feathers [whose values are present/absent] and hands [whose values are present/absent]. So we have four features whose values are **discrete**.

In the traditional programs, the above if-else-if code is hand written. Efforts put by a human being in identifying the rules and writing this piece of code where there are four features and one input are relatively less.

But could you imagine the efforts required if the numbers of features are in hundreds or thousands. It becomes a tedious job with nearly impossible timelines. Decision Tree could learn these rules from the training data. Despite other classifiers like Naive Bayes Classifier or other linear classifiers, Decision Tree could capture the non-linearity of a feature or any relation between two or more features.

Regarding the capturing relation among features in the above example, the features (wings and feathers) are co-related. For the considered example (or data set), their values are related in a way **such that** their collective value is deciding on the decision flow.

In machine learning, input dataset for the Decision Tree algorithm would be the list of feature values with the corresponding categorical value. A sample of the dataset is as shown in the Table 5-1.

Table 5-1  A sample of the dataset

| Input | Output | Features | | | |
|---|---|---|---|---|---|
| living-being | category | wings | hands | feathers | displacement |
| Joe | human | absent | present | absent | present |
| Parrot | bird | present | absent | present | present |
| Jean | human | absent | present | absent | present |
| Hibiscus | plant | absent | absent | absent | absent |
| Eagle | bird | present | absent | present | present |
| Rose | plant | absent | absent | absent | absent |

Each row in the Table 5-1 represents an observation/experiment.

In practical scenarios, the number of features could be from single digit number to thousands, and the data set would contain single digit number to millions of **entries/observations/experiments**.

The common way to build a Decision Tree is to use a greedy approach. Consider you are greedy on the number of Decision Nodes. The number of Decision Nodes should be minimal. By testing a feature value, the Dataset is broken into sub-Datasets, with a condition that the split gives maximum benefit to the classification i.e., the feature value considered (among all the possible feature value combinations) is the best available to categorize the given data set into two subsets. In each sub-Dataset, a new feature value combination is chosen, as in the former split, to divide it into smaller sub-Datasets, with the same condition that the split gives maximum benefit to the classification. The process is repeated until a Decision Node is not required to further split the sub-Dataset, and almost all of the samples in that sub-Dataset belong to a single category.

The graphical representation of Decision Tree for the Dataset mentioned above would be as shown in the Figure 5-1.

From the Figure 5-1, it is evident that the Decision Tree has made use of only two features [displacement, wings] as the other two features are redundant. Thus it needs to reduce the number of Decision Nodes.

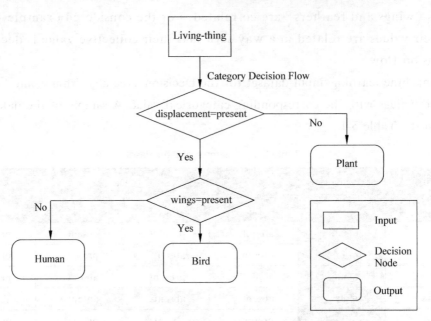

Figure 5-1  Flowchart representation of Decision Tree

  **Words**

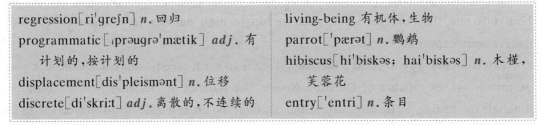

regression [riˈgreʃn] n. 回归
programmatic [ˌprəʊgrəˈmætɪk] adj. 有计划的，按计划的
displacement [dɪsˈpleɪsmənt] n. 位移
discrete [dɪˈskriːt] adj. 离散的，不连续的
living-being 有机体，生物
parrot [ˈpærət] n. 鹦鹉
hibiscus [hɪˈbɪskəs; haɪˈbɪskəs] n. 木槿，芙蓉花
entry [ˈentri] n. 条目

  **Phrases**

such that  如此……以致

  **Exercises**

I. Read the following statements carefully, and decide whether they are true (T) or false (F) according to the text.

_____ 1. The common way to build a Decision Tree is to use a SVM approach.

_____ 2. Regarding machine learning, input dataset for the Decision Tree algorithm would be the list of feature values with the corresponding categorical value.

_____ 3. Decision Tree could take the non-linearity of a feature or any relation between two or more features.

____ 4. In the Figure 5-1, the Decision Tree has used only two features [displacement, feathers].

____ 5. Decision Tree in Machine Learning is used for unsupervised learning.

II. **Choose the best answer to each of the following questions according to the text.**

1. Which of the following is right about the Decision Tree? (　　)
   A. Decision Tree could take the linearity of a feature or any relation between two or more features.
   B. Decision Tree exploits correlation between features and non-linearity in the features.
   C. Decision Tree in Machine Learning is used for unsupervised learning.
   D. The common way to build a Decision Tree is to use a SVM approach.

2. How many features are mentioned in the Figure 5-1? (　　)
   A. One
   B. Two
   C. Three
   D. Four

3. Which of the two features has Decision Tree used in the Figure 5-1? (　　)
   A. [displacement, feathers]
   B. [displacement, hands]
   C. [displacement, wings]
   D. None of the above

III. **Fill in the numbered spaces with the words or phrases chosen from the box. Change the forms where necessary.**

```
variable  independent  can  define  call
many  dependent  estimate  what  common
```

**Linear Regression**

Linear regression is a basic and ____1____ used type of predictive analysis. The overall idea of regression is to examine two things: (1) does a set of predictor variables do a good job in predicting an outcome (dependent) variable? (2) Which ____2____ in particular are significant predictors of the outcome variable, and in ____3____ way do they—indicated by the magnitude and sign of the beta estimates—impact the outcome variable? These regression ____4____ are used to explain the relationship between one dependent variable and one or more ____5____ variables. The simplest form of the regression equation with one dependent and one independent variable is ____6____ by the formula y = c + b * x, where y = estimated ____7____ variable score, c = constant, b = regression coefficient, and x = score on the independent variable.

There are ___8___ names for a regression's dependent variable. It may be ___9___ an outcome variable, criterion variable, endogenous variable, or regressand. The independent variables ___10___ be called exogenous variables, predictor variables, or regressors.

**IV. Translate the following passage into Chinese.**

**Support Vector Machine (SVM)**

A support vector machine is a supervised learning algorithm that sorts data into two categories. It is trained with a series of data already classified into two categories, building the model as it is initially trained. The task of an SVM algorithm is to determine which category a new data point belongs in. This makes SVM a kind of non-binary linear classifier.

An SVM algorithm should not only place objects into categories, but have the margins between them on a graph as wide as possible.

## Section B: K-means Clustering Algorithm and Example

"I'm clueless"

You say, looking at **an ocean of** unlabeled data, waving in front of you. It is true that the lack of labels can sometimes **freak us out**, leaving us wondering how to group the data together. But luckily, k-means **clustering** algorithm is here to rescue, one of the simplest algorithms for unsupervised clustering (dealing with data without defined categories). Assigning data points into k **clusters** based on the minimum distance, k-means clustering is simple, helpful, and effective for finding the **latent** structure in the data.

Here we provide some basic knowledge about k-means clustering algorithm and an illustrative example to help you clearly understand what it is.

K-means clustering algorithm is an unsupervised machine learning algorithm for determining which group a certain object really belongs to. What it means by "being unsupervised" is that there are no **prescribed** labels in the data denoting its structure. The main idea is to assign each observation into the cluster with the nearest **mean** (**centroid**[1]), **serving as** a prototype of the cluster.

Here are five simple steps for the k-means clustering algorithm and an example for illustration:

• Step 1: Visualize n data points and decide the number of clusters (k). Choose k random points on the graph as the centroids of each cluster. For this example, we would like to divide the data into 4 clusters, so we pick 4 random centroids (Figure 5-2).

• Step 2: Calculate the **Euclidean** distance between each data point and chosen clusters' centroids. A point is considered to be in a particular cluster if it is closer to that cluster's centroid than any other ones (Figure 5-3).

• Step 3: After assigning all **observations** to the clusters, calculate the clustering score, by **summing up** all the Euclidean distances between each data point and the corresponding centroid.

Unit 5　Machine Learning

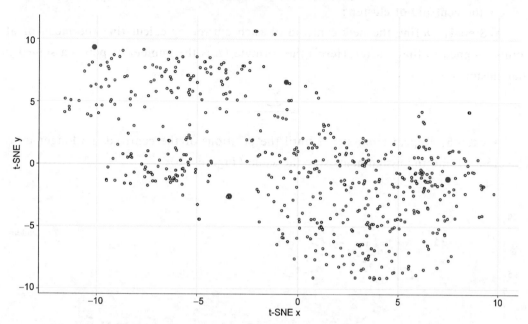

Figure 5-2　Visualize the data and pick the random centroids (which is 4 in this example)

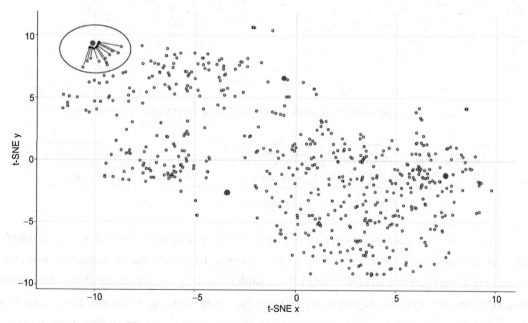

Figure 5-3　Assign each point into the cluster with the nearest centroid

$$\text{Total distances} = \sum_{j=1}^{k} \sum_{i=1}^{n} \parallel x_i^{(j)} - c_j \parallel^2$$

Where:

　　k: the number of clusters

$n_j$: the number of points belonging to cluster j

$c_j$: the centroid of cluster j

• Step 4: Define the new centroid of each cluster by calculating the mean of all points assigned to that cluster. Here's the formula (n is the number of points assigned to that cluster):

$$= \frac{\sum_{i=1}^{n} x_i}{n}$$

• Step 5: Repeat from step 2 until the positions of the centroids no longer move (Figure 5-4) and the assignments stay the same (Figure 5-5).

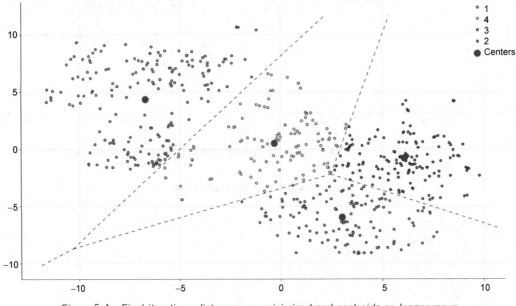

Figure 5-4  Final iteration: distances are minimized and centroids no longer move

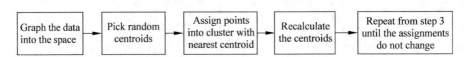

Figure 5-5  Flow chart of k-means clustering algorithm

There you go: data points are now grouped into 4 different clusters. Using a simple idea of minimizing distances between data points to group them together, k-means clustering algorithm is extremely helpful for understanding the structure of the data, how observations are classified, and interpreting the story behind. K-means clustering has been widely used in data analysis, especially in life sciences, in analyzing thousands to millions of data points in single-cell **RNA-seq** and bulk RNA-seq experiments.

Note that the Euclidean **metric** measures the distance based on the vector connecting two points, and will cause some biases for data with different scales. For example, in RNA-seq data, **gene** expression values can range from as little as 0.001 to a thousand, stretching the data points along an **axis**. That is, the variable with the smaller scale will be easily **dominated** and play little in the **convergence**, as clusters will scatter along an axis only. For this reason, it is necessary to make sure that the variables are at the same scale before using k-means clustering.

Note that before determining the number of clusters to assign the data into (the variable k), you should have an overview of the data and on what basis you want to group them. You can even apply a hierarchical clustering on the data first to briefly understand the structure of the data before choosing k by hand.

A well-known method to validate the number of clusters is the Elbow method [2], that is to run k-means clustering several times for a range of values of k (usually from 2 to 10) and pick out the value of k that causes sudden drop in the sum of **squared distances**. More specifically, for each value of k, we calculate the sum of squared distances (between each point and the corresponding centroid) and graph the results on a line chart. Choose the value where the sum of squares drops, giving an angle in the graph (a. k. a. an elbow)—that is the optimal value of k (Figure 5-6).

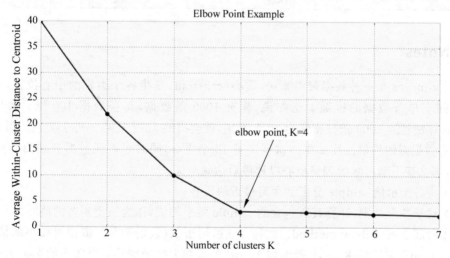

Figure 5-6　Elbow point example

  **Words**

| | |
|---|---|
| clustering['klʌstəriŋ] *n.* 聚类 | 何学的，欧几里得的 |
| cluster['klʌstə(r)] *n.* 群集，簇，集群 | observation[ˌɒbzə'veiʃn] *n.* 数据点 |
| latent['leitnt] *adj.* 潜在的，潜伏的 | RNA-seq　转录组测序技术（RNA |
| prescribe[pri'skraib] *v.* 规定 | 　　sequencing） |
| mean[miːn] *n.* 平均数，平均值 | dominate['dɒmineit] *v.* 支配，控制 |
| centroid['sentrɔid] *n.* 质心，形心 | convergence[kən'vɜːdʒəns] *n.* 趋同，融 |
| Euclidean[juː'klidiən] *adj.* 欧几里得几 | 　　合，一体化 |

metric['metrik] n. 度量标准  　　axis['æksis] n. 轴,轴线
gene[dʒi:n] n. 基因

 **Phrases**

an ocean of　极多的,无穷无尽的
freak out　崩溃,使处于极度兴奋中
serve as　用作,充当
sum up　计算……的总数
squared distances　距离平方

 **Abbreviations**

a.k.a. 亦称,又名(also known as)

 **Notes**

[1] k-means 是一种数据聚类算法,质心(centroid)是指各个类别的中心位置,质心的维数等同于单条数据的维数。比如说,你有 1000 条数据,每条数据 100 维。如果使用 k-means 算法将这 1000 条数据聚为 10 个类别,就会得到 10 个质心。每个类别的质心是该类别所有数据点的均值。比如第一次确定了 10 个质心,同时也将元数据分别归类到这 10 个质心,那么接下来可继续调整质心以致最后达到最优:

(1) 将各个示例 sample 分配到距离最近的质心;

(2) 对于各个类别,计算其所包含的 sample 的平均值,作为该类别新的质心。

[2] 肘部法则(Elbow method),此种方法适用于 K(簇的数量)值相对较小的情况,当选择的 k 值小于真正的 K 时,k 每增加 1,cost 值就会大幅地减小;当选择的 k 值大于真正的 K 时,k 每增加 1,cost 值的变化就不会那么明显。这样,正确的 k 值就会在这个转折点,类似 elbow 的地方。

 **Exercises**

I. Read the following statements carefully, and decide whether they are true (T) or false (F) according to the text.

　　____ 1. K-means clustering algorithm is a supervised machine learning algorithm.

　　____ 2. The main concept of k-means is to assign each observation into the cluster with the nearest mean (centroid), serving as a prototype of the cluster.

　　____ 3. To find the latent structure in the data k-means clustering is a simple way to assign data points into k clusters based on the minimum distance.

____ 4. "Being unsupervised" is that there are some prescribed labels in the data denoting its structure.

____ 5. Elbow method is a well-known method which validates the number of clusters.

## II. Choose the best answer to each of the following questions according to the text.

1. Which of the following is not mentioned in the text? (   )
   A. ID3
   B. Centroid
   C. Euclidean
   D. K-means

2. How many steps are mentioned for the k-means clustering algorithm and an example for illustration? (   )
   A. Two
   B. Three
   C. Four
   D. Five

3. Which of the following is right? (   )
   A. The main concept of k-means is to assign each observation into the cluster with the nearest mean (centroid), serving as a prototype of the cluster.
   B. To find the latent structure in the data k-means clustering is a simple way to assign data points into k clusters based on the minimum distance.
   C. Elbow method is a well-known method which validates the number of clusters.
   D. All of the above

## III. Fill in the numbered spaces with the words or phrases chosen from the box. Change the forms where necessary.

> understand   labor   deal   advantage   like
> base   method   as   reflect   use

**Clustering Algorithms**

Clustering algorithms can automatically recognize the pattern inside the data so __1__ to analyze the collected data without their labels. Using this advantage, three clustering-based fault diagnosis methods are presented to __2__ with some diagnosis cases of rotating machinery in which the labeled data are limited. In the first method, compensation distance evaluation technique and the weight K nearest neighbor are __3__ to recognize the mechanical faults, harnessing the merits that the computation of feature weights is simpler and the weights are easier to __4__. The second method is presented __5__ on weight fuzzy c-means, which is robust to the local structure of the data and __6__ the level of uncertainty over the most appropriate assignment.

Finally, a Hybrid clustering algorithm-based fault diagnosis ___7___ is introduced, considering the problems ___8___ the sample influence for clustering and the automatic setting of the cluster number. The results of the diagnosis cases verify that these diagnosis methods take full ___9___ of unlabeled data and reduce the human ___10___ in fault diagnosis.

## IV. Translate the following passage into Chinese.

**Ensemble Learning**

Many ensemble learning tools can be trained to produce various results. Individual algorithms may be stacked on top of each other, or rely on a "bucket of models" method of evaluating multiple methods for one system. In some cases, multiple data sets are aggregated and combined. For example, a geographic research program may use multiple methods to assess the prevalence of items in a geographic space. One of the issues with this type of research involves making sure that various models are independent, and that the combination of data is practical and works in a particular scenario.

Ensemble learning methods are included in different types of statistical software packages. Some experts describe ensemble learning as "crowdsourcing" of data aggregation.

# Part 2

# Simulated Writing: Developing Reports and Proposals (I)

报告和提案是在工作中最常写的长文档。这两者都回答了某个主题或项目的问题，或者针对某个问题提供解决方案。读者将会研究作者的报告，并且运用其中的结论和分析来帮助他们进行决策。除了商业企业之外，非盈利机构和政府机构也会撰写报告来总结或者分析研究状况。有时，组织会雇佣专业的撰稿人撰写提案以赢得合同，或获得销售机会。学会写作这些重要的文档是一项很有价值的专业技能。

**1. 了解报告和提案**

报告是一种针对特定主题交流信息而设计的书面文档。虽然有些报告可以包含分析或建议，但撰写的报告往往很客观。提案与报告很相似，但其目的在于说服和通知。提案提供了有关产品、服务或者想法的信息，并且试图说服读者接纳所建议的解决方案。报告与提案的一个关键区别在于它们被写作的时间。提案通常在制定决策过程的早期进行，此时它能够影响决策。报告通常在已经采取一些行动之后撰写。当一项活动或项目发生的时候，一些报告可以记录它们的状态。当活动或项目完结时，可以撰写其他的报告。报告和提案的类型参见图 5-7。

在开始撰写报告或提案前，请回答下面的问题：

撰写的目的是什么？

撰写报告的第一步是明确地定义目的。首先分析想要达到的目标，目标是通知、更新、分析，还是说服？目标将帮助决定应该使用的形式。

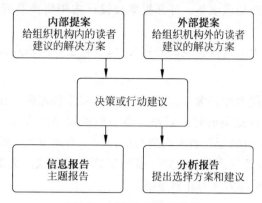

图 5-7　报告和提案的类型

**读者是谁？**

与其他类型的文档相同,撰写报告或提案时,要考虑读者。为了更好地满足读者的需求,要辨别他们理解报告或提案主旨的程度。他们想要通过阅读报告或提案了解什么？他们有可能怎样阅读？应该怎样撰写才能使信息清晰,并且使读者易懂？一定要考虑主要读者和次要读者,以及包括那些可能会阅读该文档的任何人。

**应该撰写报告还是提案？**

撰写报告是为了与他人分享信息。撰写提案是为了说服读者采纳想法、产品或者解决方案。这两者与分析报告很类似,但区别是只是这里只呈现一个建议。表 5-2 给出了何时应该撰写报告或提案的建议。

表 5-2　何时撰写报告或提案

| 场　　景 | 报告 | 提案 | 其他 |
|---|---|---|---|
| 参加一场贸易展示会,希望通告本公司的竞争对手的产品 | √ | | |
| 需要为公司流程撰写文档 | √ | | |
| 分析是购买新的计算机设备还是升级现有设备 | √ | | |
| 提议购买新的计算机设备 | | √ | |
| 为规划职员资源提议一种新方案 | | √ | |
| 为个人或组织提供公司的服务 | | √ | |
| 在所参加的一场会议上为之后的查阅总结所做的笔记 | | | 非正式笔记或大纲 |
| 为一般的受众推销公司的服务 | | | 广告 |
| 为潜在的顾客描述公司产品,并且提供样品 | | | 展示 |

**报告中会展示信息还是分析话题？**

报告可以是下述两种类型中的一个。信息报告以清晰、客观的形式展示信息。当想为读者书面总结针对某个主题的信息时,使用信息报告比较合适。意见和建议不应写在一个信息报告之中。分析报告一般会呈现数据、分析和结论。分析报告通常会提供不同的选择,鉴别优劣以得到替代方案,以及包含具体的建议。

**提案是为内部还是外部的读者而撰写？**

提案也有两种类型。内部提案建议如何在一个组织内解决问题,例如,通过改变一个程序或者使用商家的不同产品或服务。外部提案被设计来销售产品或服务于客户,并且通常为响应请求而撰写。

回答这些问题有助于决定报告应该有多长,包含什么样的信息,以及适当形式。

**2. 规划报告或提案**

有条理地组织业务报告和提案,以便使信息容易阅读和理解。在写第一句话之前,就应该有针对如何组织报告或提案的好的思路。将一般的思路组合在一起,并遵循逻辑顺序。该顺序能够满足目的,并有助于读者明白所写的内容。有逻辑地组织信息的方式应依时间、重要性以及类别,例如位置或产品来决定。撰写正式或非正式的大纲可以有助于规划有效的报告。表5-3总结了撰写大纲的注意事项。

表5-3 撰写大纲的注意事项

| 要素 | 适合提到 | 尽量避免 |
| --- | --- | --- |
| 主要思路 | · 以头脑风暴开始,列出想要包含的所有思路<br>· 选择一个作为主要的思路<br>· 写在大纲开头 | · 保留所有的思路,而不是只保留那些服务于报告或者提案的目的和那些服务于读者的思路<br>· 表述主要思路超过两句 |
| 大标题和章节 | · 选择议题并写出相应的标题<br>· 使用标准标题,例如介绍(Introduction)和结论(Conclusion)<br>· 按逻辑顺序列出标题<br>· 在正式大纲中,使用罗马数字标注大标题 | · 偏离如下的标准模式:(1)介绍(Introduction);(2)事实和发现(Facts or findings);(3)结论(Conclusion)<br>· 包含没有足够细节和证据的议题 |
| 子标题 | · 用子标题将大议题分解为子议题<br>· 以逻辑顺序列出子议题,例如,时间、重要性,或者类别<br>· 在正式大纲中,第一级子标题使用大写字母,下一级使用数字,最后一级使用小写字母 | · 以任意顺序列出子议题<br>· 使用难以解释的子标题 |

1)首先确定主要的思路

开始撰写大纲可以通过在页面顶端用一两个句子描述主要思路来开始。如果主要的思路太长,可以精简所写的内容。在页面的上方说明主要的思路,有助于在制定大纲的其余部分时专注于自己的目标。许多报告和提案的主要思路是要描述一个解决问题的办法。

2)为重要的思路使用标题

复查报告的思路和主题,并选择最重要的部分。这些都应作为大纲的主要标题。这些标题要按照逻辑顺序列出,比如从最重要的到最不重要的,或按时间顺序(如果报告强调了时间)。这些标题将成为报告的主要部分。图5-8展示了正式和非正式大纲中的标题,包括使用罗马数字、大写字母、数字和小写字母的规范。

3)为子议题创建子标题

可以将每一个主要议题分为几个思路,以便详细地讨论它们。在大纲中列出这些思路将其作为子标题。可以为每个大标题提供两个或两个以上的子标题,如图5-8所示。如果正在写一个很长的或者很复杂的报告,可以将子议题分解为更小的部分。

# Unit 5　Machine Learning

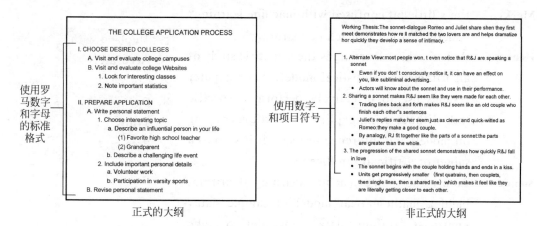

图 5-8　正式和非正式的大纲

4）将合适的章节添加进来

大多数报告和提案包括标准章节，如介绍、背景、现状、事实、提出的解决方案、总结、结论、建议、利弊、参考清单和附录。选择能够服务于报告或提案目的章节。

5）复查大纲

复查大纲的完整草案以便回答以下问题：思路是否按照逻辑顺序安排？如果大声读大纲给自己听，听起来是否有意义？标题和子标题是否具有逻辑性和平衡性？它们的重要性是否差不多？如果有必要则重新排列顺序。议题是否已经有了足够的细节或证据来支持主要的思路？如果不是，那就应该将它们添加到大纲中或者重组大纲。

转 109 页

# Part 3

# Listening & Speaking

## Dialogue：Machine Learning

在线音频

(*Before the first lesson of Machine Learning*，Mark met with Henry and Sophie in front of their classroom)

Mark：　　Excuse me，Henry and Sophie. Could you help me?[1]

Henry：　　Sure. What's the problem?

[1] Replace with：
1. Can you give me a hand?
2. Could you please do me a favor?
3. Could you do me favor?

Mark: I'm a little bit confused with machine learning? Exactly [2] what is machine learning?

Henry: Well, machine learning is the scientific study of algorithms and statistical models that computer systems use to effectively perform a specific task without using explicit instructions, relying on models and inference instead. It is seen as a subset of artificial intelligence.

Sophie: To my knowledge, machine learning algorithms build a mathematical model of sample data, known as "training data", in order to make predictions or decisions without being explicitly programmed to perform the task. Machine learning algorithms are used in the applications of email filtering, detection of network intruders, and computer vision, where it is **infeasible** to develop an algorithm of specific instructions for performing the task.

Mark: Does machine learning have some relationships with other areas?

Henry: Of course. Machine learning is closely related to computational statistics, which focuses on making predictions using computers. The study of mathematical optimization delivers methods, theory and application domains to the field of machine learning. Data mining is a field of study within machine learning, and focuses on exploratory data analysis through unsupervised learning. In its application across business problems, machine learning is also referred to as predictive analytics.

Mark: And are there any classifications for the machine learning?

Sophie: Absolutely. Machine learning tasks are classified into several broad categories. In supervised learning, the algorithm builds a mathematical model of a set of data that contains both the inputs and the desired outputs.

[2] Replace with:
1. Accurately
2. Correctly
3. Definitely
4. Truly
5. Precisely

For example, if the task were determining whether an image contained a certain object, the training data for a supervised learning algorithm would include images with and without that object (the input), and each image would have a label (the output) **designating** whether it contained the object.

Henry: In special cases, the input may be only partially available, or restricted to special feedback. Semi-supervised learning algorithms develop mathematical models from incomplete training data, where a portion of the sample inputs are missing the desired output.

Mark: Could you please name a few algorithms for supervised learning?

Henry: Sure. Classification algorithms and regression algorithms are types of supervised learning. Classification algorithms are used when the outputs are restricted to a limited set of values. For a classification algorithm that filters emails, the input would be an incoming email, and the output would be the name of the folder in which to **file** the email. For an algorithm that identifies spam emails, the output would be the prediction of either "spam" or "not spam", represented by the Boolean values true and false.

Sophie: And regression algorithms are named for their continuous outputs, meaning they may have any value within a range. Examples of a continuous value are the temperature, length, or price of an object.

Mark: So, how about unsupervised learning?

Sophie: Well, in unsupervised learning, the algorithm builds a mathematical model of a set of data which contains only inputs and no desired outputs. Unsupervised learning algorithms are used to find structure in the data, like grouping or clustering of data points.

Henry: Moreover, unsupervised learning can discover patterns in the data, and can group the inputs into categories, as in feature learning. Dimensionality reduction is the process of reducing the number of "features", or inputs, in a set of data.

Mark: OK, so what else?

Henry: Well, **reinforcement learning** algorithms are given feedback in the form of positive or negative reinforcement in a dynamic environment, and are used in autonomous vehicles or in learning to play a game against a human opponent.

Sophie: And other specialized algorithms in machine learning include topic modeling, where the computer program is given a set of natural language documents and finds other documents that cover similar topics. Machine learning algorithms can be used to find the unobservable probability density function in density estimation problems, **so on and so forth**.

Mark: So much knowledge I'm interested in! Thank you very much!

## Exercises

Work in a group, and make up a similar conversation by replacing the statements with other expressions on the right side.

## Words

| infeasible[inˈfiːzɪb(ə)l] *adj.* 不可行的，不可实行的 | designate[ˈdezɪɡneɪt] *v.* 指定，指派 file[faɪl] *v.* 把……归档 |

## Phrases

| reinforcement learning 强化学习 so on and so forth 等等 |

## Unit 5　Machine Learning

## Listening Comprehension: Supervised Learning

在线音频

*Listen to the article and answer the following 3 questions based on it. After you hear a question, there will be a break of 15 seconds. During the break, you will decide which one is the best answer among the four choices marked （A）,（B）,（C） and （D）.*

**Questions**

1. Which of the following is right?（　　）
   - （A） Supervised learning is the machine learning task of learning a function that maps an input to an output based on example input-output pairs.
   - （B） Supervised learning infers a function from labeled training data consisting of a set of training examples.
   - （C） A supervised learning algorithm analyzes the training data and produces an inferred function.
   - （D） All of the above

2. Regarding the hand-written digit recognition problem, which of the following is right?（　　）
   - （A） A reasonable data set for this problem is a collection of images of hand-written digits.
   - （B） A reasonable data set for this problem is for each image, what the digit actually is.
   - （C） A set of examples of the form（image,digit）should be considered.
   - （D） All of the above

3. Which of the following can't supervised learning do?（　　）
   - （A） Supervised learning is the machine learning task of learning a function that maps an output to an input based on example output-input pairs.
   - （B） Supervised learning is the machine learning task of learning a function that maps an input to an output based on example input-output pairs.
   - （C） Supervised learning infers a function from labeled training data consisting of a set of training examples.
   - （D） A supervised learning algorithm analyzes the training data and produces an inferred function.

 **Words**

| | |
|---|---|
| map[mæp] v. 映射 <br> entirety[in'taiərəti] n. 全部,完全 | outset['autset] n. 开始,开端 |

097

# Dictation: Unsupervised Learning

*This article will be played three times. Listen carefully, and fill in the numbered spaces with the appropriate words you have heard.*

Unsupervised learning is a ___1___ of machine learning that learns from test data that has not been ___2___, classified or categorized. Instead of ___3___ to feedback, unsupervised learning identifies **commonalities** in the data and reacts based on the presence or ___4___ of such commonalities in each new piece of data. ___5___ include supervised learning and reinforcement learning.

In the unsupervised ___6___, the training data does not contain any output information at all. We are just given input examples $X_1, \cdots, X_N$. You may wonder how we could possibly learn anything from mere inputs. Consider the coin ___7___ problem. Suppose that we didn't know the **denomination** of any of the ___8___ in the data set.

We still get similar ___9___, but they are now ___10___ so all points have the same "color". The decision regions in unsupervised learning may be ___11___ to those in supervised learning, but without the labels. However, the correct clustering is less ___12___ now, and even the number of clusters may be ___13___.

___14___, this example shows that we can learn something from the inputs by themselves. Unsupervised learning can be ___15___ as the task of **spontaneously** finding ___16___ and structure in input data. For instance, if our task is to ___17___ a set of books into topics, and we only use ___18___ properties of the ___19___ books, we can identify books that have similar ___20___ and put them together in one category, without naming that category.

## ▶ Words

| | |
|---|---|
| commonality[kɔməˈnæliti] *n.* 公共,共性 | spontaneously[spɔnˈteɪniəsli] *adv.* 自发地,自然地 |
| denomination[diˌnɔmiˈneiʃn] *n.* 面额 | |

# Unit 6

## Artificial Neural Networks

# Part 1

# Reading & Translating

## Section A: Artificial Neural Networks

With all the progress that has been made in artificial intelligence, many problems in the field continue to **tax** the abilities of computers using traditional algorithmic approaches. Sequences of instructions do not seem capable of perceiving and reasoning at levels comparable to those of the human mind. For this reason, many researchers are turning to approaches that **leverage** phenomena observed in nature. One such approach is genetic algorithms. Another approach is the artificial neural network.

Artificial neural networks provide a computer processing model that mimics networks of **neurons** in living biological systems. A biological neuron is a single cell with input **tentacles** called **dendrites** and an output tentacle called the **axon** (Figure 6-1). The signals transmitted via a cell's axon reflect whether the cell is in an inhibited or excited state. This state is determined by the combination of signals received by the cell's dendrites. These dendrites pick up signals from the axons of other cells across small gaps known as **synapses**. Research suggests that the conductivity across a single synapse is controlled by the **chemical composition of** the synapse. That is, whether the particular input signal will have an exciting or inhibiting effect on the neuron is determined by the chemical composition of the synapse. Thus it is believed that a biological neural network learns by adjusting these chemical connections between neurons.

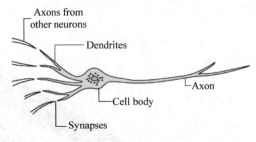

Figure 6-1　A neuron in a living biological system

A neuron in an artificial neural network is a software unit that mimics this basic understanding of a biological neuron. It produces an output of 1 or 0, depending on whether its effective input exceeds a given value, which is called the neuron's threshold value. This effective input is a **weighted** sum of the actual inputs, as represented in Figure 6-2. In this figure, a neuron is represented with an **oval** and connections between neurons are represented with arrows. The values obtained from the axons of other neurons (denoted by $v_1$, $v_2$ and $v_3$) are used as inputs to the depicted neuron. In addition to these values, each connection is associated with a weight (denoted by $w_1$, $w_2$, and $w_3$). The neuron receiving these input values multiplies each by the associated weight for the connection and then adds these products to form the effective input ($v_1 w_1 + v_2 w_2 + v_3 w_3$). If this sum exceeds the neuron's threshold value, the neuron produces an output of 1 (simulating an excited state); otherwise, the neuron produces a 0 as its output (simulating an inhibited state).

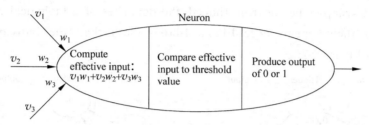

Figure 6-2   The activities within a neuron

Following the lead of Figure 6-2, we adopt the **convention** of representing neurons as circles. Where each input connects to a neuron, we record the weight associated with that input. Finally, we write the neuron's threshold value in the middle of the circle. As an example, Figure 6-3 represents a neuron with a threshold value of 1.5 and weights of $-2, 3$, and $-1$ associated with each of its input connections. Therefore, if the neuron receives the inputs 1,1, and 0, its effective input is $(1)(-2) + (1)(3) + (0)(-1) = 1$, and thus its output is 0. But, if the neuron receives 0,1, and 1, its effective input is $(0)(-2) + (1)(3) + (1)(-1) = 2$, which exceeds the threshold value. The neuron's output will thus be 1.

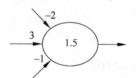

Figure 6-3   Representation of a neuron

The fact that a weight can be positive or negative means that the corresponding input can have either an inhibiting or exciting effect on the receiving neuron. (If the weight is negative, then a 1 at that input position reduces the weighted sum and thus tends to hold the effective input below the threshold value. In contrast, a positive weight causes the associated input to have an increasing effect on the weighted sum and thus increase the chances of that sum exceeding the threshold value.) Moreover, the actual size of the weight controls the degree to which the corresponding input is allowed to inhibit or excite the receiving neuron. Consequently, by adjusting the values of the weights throughout an artificial neural network, we can program the network to respond to different inputs in a predetermined manner.

Artificial neural networks are typically arranged in a topology of several layers. The input neurons are in the first layer and the output neurons are in the last. Additional layers of neurons (called hidden layers) may be included between the input and output layers. Each neuron of one layer is interconnected with every neuron in the subsequent layer. As an example, the simple network presented in Figure 6-4a is programmed to produce an output of 1 if its two inputs differ and an output of 0 otherwise. If, however, we change the weights to those shown in Figure 6-4b, we obtain a network that responds with a 1 if both of its inputs are 1s and with a 0 otherwise.

We should note that the network configuration in Figure 6-4 is far more **simplistic** than an actual biological network. A human brain contains approximately 1 011 neurons with about 104 synapses per neuron. Indeed, the dendrites of a biological neuron are so numerous that they appear more like a **fibrous** mesh than the individual tentacles represented in Figure 6-1.

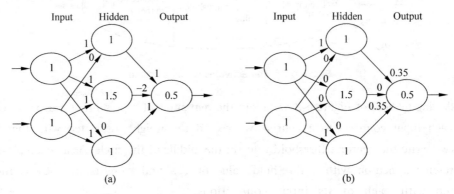

Figure 6-4  A neural network with two different programs

 **Words**

tax[tæks] *v.* 使……负重担  
leverage['li:vəridʒ] *v.* 利用  
neuron['njuərɔn] *n.* 神经元  
tentacle['tentəkl] *n.* 触角  
dendrite['dendrait] *n.* 树突  
axon['æksɔn] *n.* 轴突  
synapse['sainæps] *n.* 突触  

weight[weit] *v. & n.* 加权，权重  
oval['əuvl] *n.* 椭圆形  
convention[kən'venʃn] *n.* 习俗，常规，惯例  
simplistic[sim'plistik] *adj.* 过分简单化的  
fibrous['faibrəs] *adj.* 纤维的  

 **Phrases**

chemical composition    化学组成（成分）  
follow the lead    效仿，照样行事

Unit 6　Artificial Neural Networks

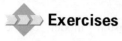 **Exercises**

**I. Read the following statements carefully, and decide whether they are true (T) or false (F) according to the text.**

　　____ 1. A biological neuron is a single cell with input dendrites called tentacles.

　　____ 2. A neuron in an artificial neural network is a software unit that simulates this basic understanding of a biological neuron.

　　____ 3. Artificial neural networks are typically arranged in a topology of only one layer.

　　____ 4. A neuron in an artificial neural network produces an output of 1 or 0, depending on whether its effective input exceeds a given value.

　　____ 5. A biological neuron is a single cell with an output axon called the tentacle.

**II. Choose the best answer to each of the following questions according to the text.**

　1. Which of the following are typically arranged in a topology of an artificial neural network? (　　)
　　A. Input layers
　　B. Output layers
　　C. Hidden layers
　　D. All of the above

　2. Which of the following is not right? (　　)
　　A. A biological neuron is a single cell with input dendrites called tentacles.
　　B. A biological neuron is a single cell with an output axon called the tentacle.
　　C. Artificial neural networks are typically arranged in a topology of only one layer.
　　D. All of the above

　3. Which of the following is right? (　　)
　　A. A neuron in an artificial neural network is a hardware unit that mimics this basic understanding of a biological neuron.
　　B. A neuron in an artificial neural network is a network unit that mimics this basic understanding of a biological neuron.
　　C. A neuron in an artificial neural network is a software unit that mimics this basic understanding of a biological neuron.
　　D. None of the above

**III. Fill in the blanks with the words or phrases chosen from the box. Change the forms where necessary.**

> discuss　interchangeable　non-linear　linear　common
> distinguish　around　correspond　various　base

103

**Activation Function**

One way to understand the activation function is to look at a visual "model" of the artificial neuron. The activation function is at the "end" of the neural structure, and ___1___ roughly to the axon of a biological neuron.

Another way to understand it is to look at the terminology ___2___ its use. IT professionals talk about the activation function when ___3___ either a binary output—either a 1 or a 0—or a function that graphs a range of outputs ___4___ on inputs. In these cases, IT professionals and others often use the terms "transfer function" and "activation function" ___5___, although the transfer function is more often associated with the graph that scans a range of outputs. ___6___ functions guide the output that filters through the layers of the neural network to the final output layer of neurons or nodes.

It is also important to ___7___ between linear and non-linear activation functions. Where ___8___ activation functions maintain a constant, ___9___ activation functions create more variation which utilizes the build of the neural network. Functions like sigmoid and ReLU are ___10___ used in neural networks to help build working models.

**IV. Translate the following passage into Chinese.**

**Non-Linear Activation Function**

Activation functions are any functions that define the output of a neuron. The activation function associated with each neurons in a neural network determines whether it should be activated or not, based on the output of that function. There are three types of activation functions—Binary, Linear and Non-Linear activation function.

Input to the neural network is usually linear transformation (i.e. input * weight + bias), but most of the real world data are non-linear. So, to make that input non-linear, non-linear activation functions are used. Non-linear Activation is the functions that add non-linearity into the network.

## Section B: Handwritten Digit Recognition

Recognizing handwritten digits is an important problem with many applications, including automated sorting of mail by postal code, automated reading of **checks** and **tax returns**, and data entry for hand-held computers. It is an area where rapid progress has been made, in part because of better learning algorithms and in part because of the availability of better training sets. The United States National Institute of Science and Technology (NIST) has **archived** a database of 60 000 labeled digits, each $20 \times 20 = 400$ **pixels** with 8-bit **grayscale** values. It has become one of the standard **benchmark** problems for comparing new learning algorithms. Some example digits are shown in Figure 6-5.

Many different learning approaches have been tried. One of the first, and probably the simplest, is the 3-nearest-neighbor classifier, which also has the advantage of

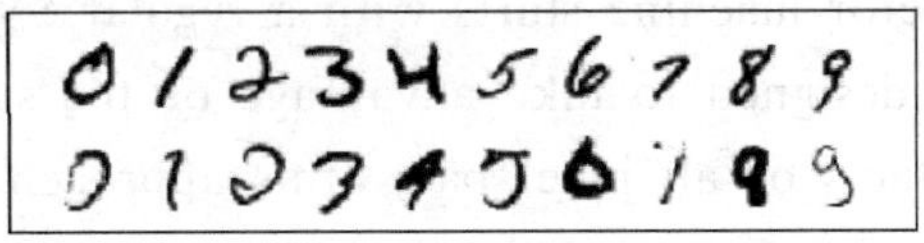

Figure 6-5  Examples from the NIST database of handwritten digits. Top row: examples of digits 0-9 that are easy to identify. Bottom row: more difficult examples of the same digits.

requiring no training time. As a memory-based algorithm, however, it must store all 60 000 images, and its run time performance is slow. It achieved a test error rate of 2.4%.

A single-hidden-layer neural network was designed for this problem with 400 input units (one per pixel) and 10 output units (one per class). Using cross-validation, it was found that roughly 300 hidden units gave the best performance. With full interconnections between layers, there were a total of 123 300 weights. This network achieved a 1.6% error rate.

A series of specialized neural networks called LeNet were devised to **take advantage of** the structure of the problem—that the input consists of pixels in a two-dimensional **array**, and that small changes in the position or **slant** of an image are unimportant. Each network had an input layer of $32 \times 32$ units, onto which the $20 \times 20$ pixels were centered so that each input unit is presented with a local neighborhood of pixels. This was followed by three layers of hidden units. Each layer consisted of several **planes** of $n \times n$ arrays, where n is smaller than the previous layer so that the network is **down-sampling** the input, and where the weights of every unit in a plane are constrained to be identical, so that the plane is acting as a feature detector: it can **pick out** a feature such as a long vertical line or a short semi-circular arc. The output layer had 10 units. Many versions of this architecture were tried; a representative one had hidden layers with 768, 192, and 30 units, respectively. The training set was augmented by applying **affine transformations** to the actual inputs: shifting, slightly rotating, and scaling the images. (Of course, the transformations have to be small, or else a 6 will be transformed into a 9!) The best error rate achieved by LeNet was 0.9%.

A boosted neural network combined three copies of the LeNet architecture, with the second one trained on a mix of patterns that the first one got 50% wrong, and the third one trained on patterns for which the first two disagreed. During testing, the three nets voted with the majority ruling. The test error rate was 0.7%.

A support vector machine with 25,000 support vectors achieved an error rate of 1.1%. This is remarkable because the SVM technique, like the simple nearest-neighbor approach, required almost no thought or iterated experimentation on the part of the developer, yet it still came close to the performance of LeNet, which had had years of development.

Indeed, the support vector machine makes no use of the structure of the problem, and would perform just as well if the pixels were presented in a **permuted** order.

A virtual support vector machine starts with a regular SVM and then improves it with a technique that is designed to take advantage of the structure of the problem. Instead of allowing products of all pixel pairs, this approach concentrates on kernels formed from pairs of nearby pixels. It also augments the training set with transformations of the examples, just as LeNet did. A virtual SVM achieved the best error rate recorded **to date**, 0.56%.

Shape matching is a technique from computer vision used to align corresponding parts of two different images of objects. The idea is to pick out a set of points in each of the two images, and then compute, for each point in the first image, which point in the second image it corresponds to. From this alignment, we then compute a transformation between the images. The transformation gives us a measure of the distance between the images. This distance measure is better motivated than just counting the number of differing pixels, and it **turns out** that a 3-nearest neighbor algorithm using this distance measure performs very well. Training on only 20 000 of the 60 000 digits, and using 100 sample points per image extracted from a Canny edge detector [1], a shape matching classifier achieved 0.63% test error.

Humans are estimated to have an error rate of about 0.2% on this problem. This figure is somewhat suspect because humans have not been tested as extensively as have machine learning algorithms. On a similar data set of digits from the United States Postal Service, human errors were at 2.5%.

## Words

check[tʃek] n. 支票
archive[ˈɑːrkaiv] v. 把……存档
pixel[ˈpiksl] n.（显示器或电视机图像的）像素
grayscale[ˈgreiskeil] n. 灰度
benchmark[ˈbentʃmɑːk] n. 基准

array[əˈrei] n. 数组
slant[slɑːnt] n. 倾斜
plane[plein] n. 平面
down-sample 降低取样
permute[pəˈmjuːt] v. 改变……的次序，重新排列

## Phrases

tax return 纳税申报单
take advantage of 利用
pick out 挑选出
affine transformation 仿射变换
to date 至今，迄今为止
turn out 结果是，事实证明

## Notes

[1] 图像的边缘检测算法(Canny Edge Detector),用于检测出图像物体的边界(boundaries)。

## Exercises

Ⅰ. **Read the following statements carefully, and decide whether they are true (T) or false (F) according to the text.**

____ 1. A support vector machine with 25 000 support vectors achieved an error rate of 0.7%.

____ 2. 3-nearest-neighbor classifier has the advantage of requiring no training time.

____ 3. Storing all 60 000 images, and its run time performance is slow, 3-nearest-neighbor classifier achieved a test error rate of 2.4%.

____ 4. Training on only 20 000 of the 60 000 digits, and using 100 sample points per image extracted from a Canny edge detector, a shape matching classifier achieved 0.9% test error.

____ 5. The United States National Institute of Science and Technology (NIST) has archived a database of 5 000 labeled digits, each 30×30 = 900 pixels with 16-bit grayscale values.

Ⅱ. **Choose the best answer to each of the following questions according to the text.**

1. Which of the following is not mentioned about the applications of recognizing handwritten digits? (     )
   A. Automated reading of checks and tax returns
   B. Data entry for hand-held computers
   C. Automated sorting of mail by postal code
   D. None of the above

2. Which of the following was achieved by a support vector machine with 25 000 support vectors? (     )
   A. An error rate of 0.7%
   B. An error rate of 0.9%
   C. An error rate of 1.1%
   D. An error rate of 2.5%

3. Which of the following is right? (     )
   A. The United States National Institute of Science and Technology (NIST) has archived a database of 5 000 labeled digits, each 30×30 = 900 pixels with 16-bit grayscale values.
   B. Training on only 20 000 of the 60 000 digits, and using 100 sample points per image extracted from a Canny edge detector, a shape matching classifier achieved 0.7% test error.
   C. A support vector machine with 25 000 support vectors achieved an error rate

of 0.56%.

D. None of the above

III. Fill in the numbered spaces with the words or phrases chosen from the box. Change the forms where necessary.

> adjust  note  similar  simple  include
> correspond  far  create  according  general

**Backpropagation**

The Backpropagation neural network is a multilayered, feedforward neural network and is by ___1___ the most extensively used. It is also considered one of the ___2___ and most general methods used for supervised training of multilayered neural networks. Backpropagation works by approximating the non-linear relationship between the input and the output by ___3___ the weight values internally. It can further be generalized for the input that is not included in the training patterns (predictive abilities).

___4___, the Backpropagation network has two stages, training and testing. During the training phase, the network is "shown" sample inputs and the correct classifications. For example, the input might be an encoded picture of a face, and the output could be represented by a code that ___5___ to the name of the person.

A further note on encoding information—a neural network, as most learning algorithms, needs to have the inputs and outputs encoded ___6___ to an arbitrary user defined scheme. The scheme will define the network architecture so that once a network is trained, the scheme cannot be changed without ___7___ a totally new net. ___8___ there are many forms of encoding the network response.

The topology of the Backpropagation neural network ___9___ an input layer, one hidden layer and an output layer. It should be ___10___ that Backpropagation neural networks can have more than one hidden layer.

IV. Translate the following passage into Chinese.

**Feedforward Neural Network**

The feedforward neural network, as a primary example of neural network design, has a limited architecture. Signals go from an input layer to additional layers. Some examples of feedforward designs are even simpler. For example, a single-layer perceptron model has only one layer, with a feedforward signal moving from a layer to an individual node. Multi-layer perceptron models, with more layers, are also feedforward.

In the days since scientists devised the first artificial neural networks, the technology world has made all sorts of progress in building more sophisticated models. There are recurrent neural networks and other designs that contain loops or cycles. There are models that involve backpropagation, where the machine learning system essentially optimizes by sending data back through a system. The feedforward neural network does not involve any of this type of design, and so it is a unique type of system that is good for learning these designs for the first time.

## Part 2

## Simulated Writing: Developing Reports and Proposals (II)

接 93 页

**3. 撰写短篇报告**

短篇报告专注于单一的思路或议题，并且不需要像较长的文档那样非常正式或注重细节。短的形式通常用于定期(或活动)的报告、进度更新、旅行报告以及其他以简洁为宜的情境。如果需要讨论这些议题中的几个，可以在单独的短篇报告中讨论它们的每一个。短篇报告不应用于重要的事项，不应用于正式的场合，以及不应用于撰写提案。

1) 使用五段式

如图 6-6 和图 6-7 所示，最简单的报告格式采用五段式来展现有关特定议题的信息：第一段提供简要概述，并指出主要思路；接下来的 3 段能够支持议题；结论段可以总结所写的内容，并重申自己的主要思路。短篇报告并不需要包括尽可能多的细节或像完整的正式报告那样涵盖尽可能多的议题。重点应放在读者想要或需要知道的方面。

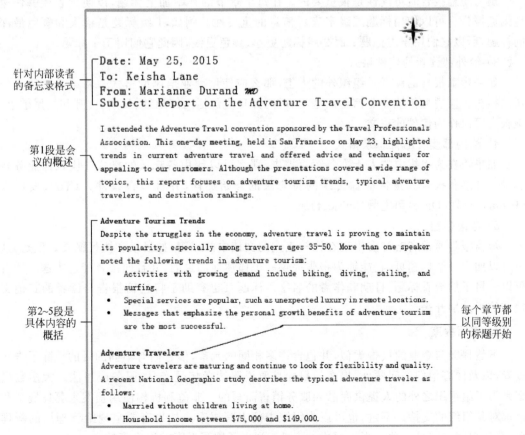

图 6-6　短篇报告的开始部分

> **Destination Trends**
> Three organizations track destination rankings. Mr. Alan Bittman, president of Travel Professionals Association, provided the following composite statistics:
> The five most popular adventure destinations in 2015 in developed countries are Switzerland, Sweden, New Zealand, United Kingdom, and Spain (in order of popularity).
> The five most popular adventure destinations in 2015 in developing countries are Estonia, Chile, Slovak Republic, Czech Republic, and Hungary (in order of popularity).
> Air travel is 20% more expensive this year than two years ago, so countries and regions that are investing in improved high-speed rail systems will be in demand.
>
> **Summary Presentation**
> This report covers the most striking findings at the conference. Many more topics were discussed that would be valuable to most people at Quest Specialty Travel. I would be happy to develop a presentation summarizing the conference and give the presentation at a staff meeting or luncheon. If this is acceptable to you, let me know when I can schedule the meeting.

(结语提供了后续的描述)

图 6-7 短篇报告的结尾

2) 以一个标题开始每个章节

撰写短篇报告时可读性是很重要的。在每个章节的开始加上标题,以便读者能够快速地浏览报告。可以使用标题层级来表示信息的重要性。例如,1级标题是最大和颜色最深的标题,所以它们用于主标题;而2级标题更小、颜色更浅,因此它们用于子标题。

3) 给外部读者的信件格式

如果短篇报告的读者是组织外的人士,那么应使用商业信函的格式。将第一页印上公司的抬头。应遵循传统商业信函的基本方针,并且应包括内部地址、日期、称呼和结尾敬语。该报告可以作为信件的正文。

4) 给内部读者的备忘录格式

如果正在为组织内的人士准备短篇报告,那么可以使用备忘录的格式。应该像业务备忘录一样准备报告。报告应像传统的备忘录的开始那样,并且包含收件人(To)、发件人(From)、日期(Date)和主题(Subject)域。

5) 考虑标题页

短篇报告通常省略许多出现在比较正式的文档中的元素,如目录和执行摘要。但是,这里可以加入一个标题页,它能够为报告提供一个描述性的封面,并且向读者传达主题。读者可以一目了然地看标题、日期和作者的名字。标题页也有助于将这篇报告与读者的其他文档和信函区别开来。

### 4. 撰写提案

虽然提案与企业的报告类似,并包含许多相同的元素,但它们有不同的目的。除了告知读者,以及回答他们的问题外,提案的目标是要说服读者接受产品、服务或想法。大多数提案是为了给组织之外的人提供商品和服务销售而写的。非盈利机构使用提案来募集资金和请求对其组织的支持。不过,也可以为组织中的人写提案。如果有要呈现给管理层的新理念,那么提案往往是一种适当的形式。表6-1列出了撰写提案的注意事项。

表 6-1　撰写提案的注意事项

| 要　素 | 适 合 提 到 | 尽 量 避 免 |
|---|---|---|
| 征求提案 | • 当有人要求撰写提案时,通常使用提案征求书(RFP)<br>• 使用与正式报告相似的格式,包括标题页、执行摘要,以及目录 | • 写不能提供的解决方案<br>• 忽视 RFP 中的要求 |
| 非征求提案 | • 当读者也许不了解所能提供的产品或服务时<br>• 使用与短篇企业报告类似的非正式格式<br>• 包括介绍、背景信息、提议的产品或解决方案、职员要求、预算和日程安排 | • 仅仅为了销售——要给出有价值的解决方案<br>• 提供虚假的收支情况——这经常成为合同中的基础 |
| 产品提案 | • 建议读者购买产品或服务<br>• 将即将销售的那些产品包含进来,提供商品的支持信息,商讨读者如何获益,以及估算成本 | • 过多提供那些不再需要的服务的细节<br>• 没有包含截止日期就估计一个预算 |
| 解决方案提案 | • 提出建议、服务,或对一个问题的复杂解法<br>• 描述接受建议的好处<br>• 为信誉提供证明 | • 提供模糊的方案,使用确定的建议<br>• 忘记提到间接或直接的收益 |

1) 撰写征求提案

当有人请求时,可以撰写征求提案,通常使用提案征求书(RFP),其中规定了提案的要求。当客户想要购买产品或服务,而通过传统渠道太复杂或订单太独特时,就可以使用 RFP。RFP 描述了客户要求的细节。

2) 撰写非征求提案

当客户没有要求所提供的产品或服务时,可以撰写非征求提案。在这种情况下,需要特别具有说服力,因为读者是不期待报价的。尽快解释所提议的,以及为什么提议对他们是有价值的,以便除了第一页之外,他们也能够继续读下去。

3) 撰写产品提案

建议读者购买产品或服务的提案被称为产品提案或商品提案。它们能够展示商品的信息,以及报出购买的价格。产品提案专注于所要提供的产品,提供有关商品的支持信息,讨论读者将如何从商品中受益,以及估算成本。

4) 撰写解决方案提案

解决方案提案(有时称为服务提案)可以建议想法、服务或复杂的解决方案。它们首先描述了问题,并且确定如何提议来解决这个问题。接下来就是制定详细计划来完成解决方案,以及要执行的日程安排表。

5) 专注于关键要素

提案的目标是要说服读者。如果没有有说服力的要素,那么所撰写的只是一个信息报告。想要具有说服力,提案应当包括解决方案、收益和信誉度,如图 6-8 所示。

6) 销售要求

读者可能会被所展示的信息激发起来,但不知道应如何去遵循提案。可以呼吁读者采取行动来结束提案,告诉他们需要做什么来接受提案,并且按照流程继续做下去。

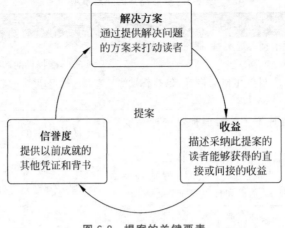

图 6-8　提案的关键要素

## Part 3

## Listening & Speaking

### Dialogue：Artificial Neural Network

（*Henry would like to learn deep learning*，*however he knows he should have some knowledge about artificial neural network*.）

**Henry**： Hi Sophie and Mark，do you know anything about artificial neural network？

**Sophie**： Sure. An artificial neural network is an interconnected group of nodes，similar to the vast network of neurons in a brain. Here，each circular node represents an artificial neuron and an arrow represents a connection from the output of one artificial neuron to the input of another.

**Mark**： To my knowledge，Artificial Neural Networks （ANN） or connectionist systems are computing systems vaguely inspired by the biological neural networks that constitute animal brains. The neural network itself is not an algorithm，but rather[1] a framework for many different machine learning algorithms to work together and process complex data inputs. Such systems "learn" to perform tasks by

[1]Replace with：
1. but instead
2. but otherwise

|  |  |
|---|---|
| | considering examples, generally without being programmed with any task-specific rules. |
| Henry: | Could you please give me an example, Mark? |
| Mark: | Of course. For example,[2] in image recognition, they might learn to identify images that contain cats by analyzing example images that have been manually labeled as "cat" or "no cat" and using the results to identify cats in other images. They do this without any prior knowledge about cats, for example, that they have fur, tails, **whiskers** and cat-like faces. Instead, they automatically generate identifying characteristics from the learning material that they process. |
| Sophie: | That is right. An ANN is based on a collection of connected units or nodes called artificial neurons, which loosely model the neurons in a biological brain. Each connection, like the synapses in a biological brain, can transmit a signal from one artificial neuron to another. An artificial neuron that receives a signal can process it and then signal additional artificial neurons connected to it. |
| Henry: | A signal? |
| Sophie: | Yes. In common ANN implementation, the signal at a connection between artificial neurons is a real number, and the output of each artificial neuron is computed by some non-linear function of the sum of its inputs. The connections between artificial neurons are called 'edges'. Artificial neurons and edges typically have a weight that adjusts as learning proceeds. The weight increases or decreases the strength of the signal at a connection. |
| Mark: | Right. Artificial neurons may have a threshold such that the signal is only sent if the **aggregate** signal crosses that threshold. Typically,[3] artificial neurons are aggregated into layers. Different layers may perform different kinds of transformations on their inputs. Signals travel from the first layer (the input layer), to the last layer (the output layer), possibly after traversing the layers multiple times. |

[2] Replace with:
1. For instance,
2. As an example,

[3] Replace with:
1. Generally,
2. Usually,

Henry: So what is the original goal of the ANN approach?

Sophie: The original goal of the ANN approach was to solve problems in the same way that a human brain would. However, over time, attention moved to performing specific tasks, leading to deviations from biology. Artificial neural networks have been used on a variety of tasks, including computer vision, speech recognition, machine translation, social network filtering, playing board and video games and medical diagnosis.

Henry: Sounds interesting. I think I should ask you more, maybe sometime later. Anyway thanks a lot.

Sophie & Mark: That's all right.

## Exercises

Work in a group, and make up a similar conversation by replacing the statements with other expressions on the right side.

## Words

| whisker['wɪskə(r)] n. 胡须 | aggregate['æɡrɪɡət] adj. 聚合的，集合的 |
|---|---|

## Listening Comprehension: Training Artificial Neural Networks

Listen to the article and answer the following 3 questions based on it. After you hear a question, there will be a break of 15 seconds. During the break, you will decide which one is the best answer among the four choices marked (A), (B), (C) and (D).

**Questions**

1. By which way does the network's performance approach the desired behavior? (  )
   (A) By programming artificial neural networks
   (B) By training artificial neural networks
   (C) By merging artificial neural networks
   (D) None of the above

2. In order to accomplish the situation in which the robot needs to learn to distinguish between walls and floor under a wide variety of lighting conditions, what could we do? (  )
   (A) We could build an artificial neural network whose inputs consist of values

indicating the color characteristics of an individual pixel in the image.

(B) We could build an artificial neural network whose inputs consist of a value indicating the overall brightness of the entire image.

(C) We could then train the network by providing it with numerous examples of pixels representing parts of walls and floors under various lighting conditions.

(D) All of the above

3. From which device does a robot receive the information when trying to understand its environment? (　　)

(A) Its photographer

(B) Its video camera

(C) Its speaker

(D) Its scanner

## Words

increment [ˈiŋkrəmənt] *n*. 增量

## Phrases

at first glance 乍一看，初看

## Dictation：Applications of Neural Networks

*This article will be played three times. Listen carefully, and fill in the numbered spaces with the appropriate words you have heard.*

There are ___1___ applications for neural networks in all areas of industry, especially for deep learning. ___2___ recognition in all of its forms is a very important area, whether analysis of photos to recognize people or faces, ___3___ of fish **swarms** in **sonar** readings, recognition and ___4___ of military ___5___ in radar ___6___, or any number of other applications. Neural networks can also be ___7___ to recognize spoken language and hand ___8___ text.

Neural networks are not only used for recognizing objects and ___9___. They can be trained to control ___10___ cars or ___11___ based on sensor data, as well as for ___12___ controlling search in **backgammon** and chess computers.

For quite some time, neural networks, in addition to ___13___ methods, have been used successfully to forecast ___14___ prices and to ___15___ the **creditworthiness** of bank customers. Speed trading of international financial transactions would be impossible without the help of smart and fast neural networks that ___16___ decide about buying or selling.

Other machine learning ____17____ can as well be used for many of these applications. Due to the great ____18____ success of data mining, decision tree learning and support ____19____ machines, there are neural algorithms for many applications as well as others that are not biologically ____20____ at all. The field of neural networks is a subarea of machine learning.

 **Words**

| | |
|---|---|
| swarm[swɔːm] n. 一大群 | n. 西洋双陆棋戏 |
| sonar['səʊnɑː(r)] n. 声呐 | creditworthiness['kredɪtwɜːrðɪnəs] n. 信 |
| backgammon['bækgæmən；ˌbækˈgæmən] | 誉，信用可靠程度 |

# Unit 7

## Deep Learning

# Part 1

# Reading & Translating

## Section A: Deep Learning, Machine Learning, and AI

Consider the following definitions to understand deep learning vs. machine learning vs. AI:

(1) Deep learning is a subset of machine learning that's based on artificial neural networks. The learning process is deep because the structure of artificial neural networks consists of multiple input, output, and hidden layers. Each layer contains units that transform the input data into information that the next layer can use for a certain predictive task. Thanks to this structure, a machine can learn through its own data processing.

(2) Machine learning is a subset of artificial intelligence that uses techniques (such as deep learning) that enable machines to use experience to **improve at** tasks. The learning process is based on the following steps:

- Feed data into an algorithm. (In this step you can provide additional information to the model, for example, by performing feature extraction.)
- Use this data to train a model.
- Test and deploy the model.
- Consume the deployed model to do an automated predictive task. (In other words, call and use the deployed model to receive the predictions returned by the model.)

(3) Artificial Intelligence (AI) is a technique that enables computers to mimic human intelligence. It includes machine learning.

It's important to understand the relationship among AI, machine learning, and deep learning (Figure 7-1). Machine learning is a way to achieve artificial intelligence. By using machine learning and deep learning techniques, you can build computer systems and applications that do tasks that are commonly associated with human intelligence. These tasks include image recognition, speech recognition, and language translation.

Now that you have the overview of machine learning vs. deep learning, let's compare the two techniques. In machine learning, the algorithm needs to be told how to make an accurate prediction by consuming more information (for example, by performing feature extraction). In deep learning, the algorithm can learn how to make an accurate prediction through its own data processing, thanks to the artificial neural network

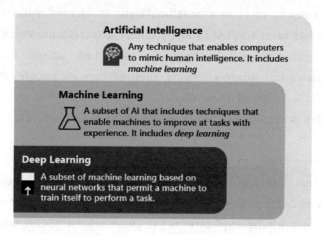

Figure 7-1　The relationship among AI, machine learning, and deep learning.

structure.

Table 7-1 compares the two techniques in more detail.

Table 7-1　Machine learning vs. deep learning

|  | All machine learning | Only deep learning |
|---|---|---|
| Number of data points | Can use small amounts of data to make predictions | Needs to use large amounts of training data to make predictions |
| Hardware dependencies | Can work on low-end machines. It doesn't need a large amount of computational power | Depends on high-end machines. It inherently does a large number of matrix multiplication operations. A **GPU** can efficiently optimize these operations |
| **Featurization** process | Requires features to be accurately identified and created by users | Learns high-level features from data and creates new features by itself |
| Learning approach | Divides the learning process into smaller steps. It then combines the results from each step into one output | Moves through the learning process by resolving the problem on an end-to-end basis |
| Execution time | Takes comparatively little time to train, ranging from a few seconds to a few hours | Usually takes a long time to train because a deep learning algorithm involves many layers |
| Output | The output is usually a numerical value, like a score or a classification | The output can have multiple formats, like a text, a score or a sound |

　　Because of the artificial neural network structure, deep learning **excels at** identifying patterns in unstructured data such as images, sound, video, and text. For this reason, deep learning is rapidly transforming many industries, including healthcare, energy, finance, and transportation. These industries are now rethinking traditional business processes.

　　Named-entity recognition is a deep learning method that takes a piece of text as input and transforms it into a pre-specified class. This new information could be a postal code, a date, a product ID. The information can then be stored in a structured **schema** to build a list of addresses or serve as a benchmark for an identity validation engine.

Deep learning has been applied in many object detection use cases. Object detection comprises two parts: image classification and then image localization. Image classification identifies the image's objects, such as cars or people. Image localization provides the specific location of these objects.

Object detection is already used in industries such as gaming, retail, tourism, and self-driving cars.

Like image recognition, in **image captioning**, for a given image, the system must generate a **caption** that describes the contents of the image. When you can detect and label objects in photographs, the next step is to turn those labels into descriptive sentences.

Usually, image captioning applications use **convolutional** neural networks to identify objects in an image and then use a **recurrent** neural network to turn the labels into consistent sentences.

Machine translation takes words or sentences from one language and automatically translates them into another language. Machine translation has been **around** for a long time, but deep learning achieves impressive results in two specific areas: automatic translation of text (and translation of speech to text) and automatic translation of images.

With the appropriate data transformation, a neural network can understand text, audio, and visual signals. Machine translation can be used to identify **snippets** of sound in larger audio files and **transcribe** the spoken word or image as text.

Text analytics based on deep learning methods involves analyzing large quantities of text data (for example, medical documents or expenses receipts), recognizing patterns, and creating organized and concise information out of it.

Companies use deep learning to perform text analysis to detect **insider trading** and compliance with government regulations. Another common example is insurance fraud: text analytics has often been used to analyze large amounts of documents to recognize the chances of an insurance **claim** being fraud.

Artificial neural networks are formed by layers of connected nodes. Deep learning models use neural networks that have a large number of layers.

The following sections explore most popular artificial neural network **typologies**.

The **feedforward** neural network is the most basic type of artificial neural network. In a feedforward network, information moves in only one direction from input layer to output layer. Feedforward neural networks transform an input by putting it through a series of hidden layers. Every layer is made up of a set of neurons, and each layer is fully connected to all neurons in the layer before. The last fully connected layer (the output layer) represents the generated predictions.

Recurrent neural networks are a widely used artificial neural network. These networks save the output of a layer and feed it back to the input layer to help predict the layer's outcome. Recurrent neural networks have great learning abilities. They're widely used for complex tasks such as time series forecasting, learning handwriting and recognizing language.

A convolutional neural network is a particularly effective artificial neural network, and it presents a unique architecture. Layers are organized in three dimensions: width, height, and depth. The neurons in one layer connect not to all the neurons in the next layer, but only to a small region of the layer's neurons. The final output is reduced to a single vector of probability scores, organized along the depth dimension.

Convolutional neural networks have been used in areas such as video recognition, image recognition and recommender systems.

## Words

| | |
|---|---|
| featurization [ˌfiːtʃəraɪˈzeɪʃən] n. 特征化，特性化 | around [əˈraʊnd] adv. 到处，大约 |
| schema [ˈskiːmə] n. 模式，计划 | snippet [ˈsnɪpɪt] n. 片段 |
| caption [ˈkæpʃn] n. 标题 | transcribe [trænˈskraɪb] v. 改编，转录，抄写 |
| convolutional [ˌkɒnvəˈluːʃən(ə)l] adj. 卷积的 | claim [kleɪm] n. (向公司等)索赔 |
| recurrent [rɪˈkʌrənt] adj. 循环的 | typology [taɪˈpɒlədʒi] n. 分类法 |
| | feedforward [ˈfiːdfɔːwəd] n. 前馈(控制) |

## Phrases

improve at　提升，改善
excel at　擅长于，擅长
image captioning　图像标注
insider trading　内线交易

## Abbreviations

GPU (Graphics Processing Unit)　图形处理器

## Exercises

I. Read the following statements carefully, and decide whether they are true (T) or false (F) according to the text.

　　____ 1. AI is a subset of machine learning.
　　____ 2. Machine learning is a subset of deep learning.

_____ 3. Object detection consists of two parts: video classification and then video localization.

_____ 4. Recurrent neural networks are widely used for complex tasks such as time series forecasting, learning handwriting and recognizing language.

_____ 5. Deep learning models use neural networks that have many layers.

II. Choose the best answer to each of the following questions according to the text.

1. Which of the following is right? (　　)
   A. Artificial Intelligence (AI) includes machine learning.
   B. Machine learning is a subset of artificial intelligence.
   C. Deep learning is a subset of machine learning.
   D. All of the above

2. Which of the following belongs to artificial neural network? (　　)
   A. Convolutional neural network
   B. Recurrent neural network
   C. Feedforward neural network
   D. All of the above

3. How many kinds of artificial neural networks are mentioned in this text? (　　)
   A. One           B. Two
   C. Three         D. Four

III. Fill in the blanks with the words or phrases chosen from the box. Change the forms where necessary.

> think　like　refer　simple　actual　input
> comprise　give　depend　connect

**Multi-Layer Perceptron**

　　A Multi-Layer Perceptron (MLP) is a deep, artificial neural network. A neural network is ___1___ of layers of nodes which activate at various levels ___2___ on the previous layer's nodes. When ___3___ about neural networks, it may be helpful to isolate your thinking to a single node in the network.

　　MLP ___4___ to a neural network with at least three layers of nodes, an input layer, some number of intermediate layers, and an output layer. Each node in a ___5___ layer is ___6___ to every node in the adjacent layers. The ___7___ layer is just that, it is the way the network takes in data. The intermediate layer(s) are the computational machine of the network, and they ___8___ transform the input to the output. The output layer is the way that results are obtained from the neural network. In a ___9___ network where the responses are binary, there would ___10___ be only one node in the output layer, which outputs a probability like in logistic regression.

122

## IV. Translate the following passage into Chinese.

**LSTM Networks**

Long Short Term Memory networks—usually just called "LSTMs"—are a special kind of RNN, capable of learning long-term dependencies. They were introduced by Hochreiter & Schmidhuber (1997), and were refined and popularized by many people in following work. They work tremendously well on a large variety of problems, and are now widely used.

LSTMs are explicitly designed to avoid the long-term dependency problem. Remembering information for long periods of time is practically their default behavior, not something they struggle to learn!

## Section B: Convolutional Neural Network

In deep learning, a Convolutional Neural Network (CNN, or ConvNet) is a class of deep neural networks, most commonly applied to analyzing visual **imagery**.

CNNs use a variation of **multilayer perceptions** designed to require minimal preprocessing. They are also known as **shift invariant** or Space Invariant Artificial Neural Networks (SIANN), based on their **shared-weights** architecture and translation invariance characteristics.

Convolutional networks were inspired by biological processes in that the connectivity pattern between neurons resembles the organization of the animal visual **cortex**. Individual **cortical** neurons respond to **stimuli** only in a restricted region of the **visual field** known as the **receptive field**. The receptive fields of different neurons partially overlap such that they cover the entire visual field.

CNNs use relatively little pre-processing compared to other image classification algorithms. This means that the network learns the filters that in traditional algorithms were **hand-engineered**. This independence from prior knowledge and human effort in feature design is a major advantage.

They have applications in image and video recognition, **recommender systems**, image classification, medical image analysis, and natural language processing.

A convolutional neural network consists of an input and an output layer, as well as multiple hidden layers (Figure 7-2). The hidden layers of a CNN typically consist of convolutional layers, ReLU [1] layer i. e. activation function, **pooling layers**, fully connected layers and normalization layers.

Description of the process as a convolution in neural networks is **by convention**. Mathematically it is a cross-correlation rather than a convolution (although cross-correlation is a related operation). This only has significance for the indices in the matrix, and thus which weights are placed at which index.

Convolutional layers apply a convolution operation to the input, passing the result to the next layer. The convolution emulates the response of an individual neuron to visual

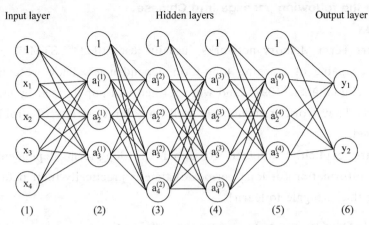

Figure 7-2  A convolutional neural network

stimuli.

Each convolutional neuron processes data only for its receptive field. Although fully connected feedforward neural networks can be used to learn features as well as classify data, it is not practical to apply this architecture to images. A very high number of neurons would be necessary, even in a shallow (opposite of deep) architecture, due to the very large input sizes associated with images, where each pixel is a relevant variable. For instance, a fully connected layer for a (small) image of size $100 \times 100$ has 10 000 weights for each neuron in the second layer. The convolution operation brings a solution to this problem as it reduces the number of free parameters, allowing the network to be deeper with fewer parameters. For instance, regardless of image size, **tiling** regions of size $5 \times 5$, each with the same shared weights, requires only 25 learnable parameters. In this way, it resolves the vanishing or exploding **gradients** problem in training traditional multi-layer neural networks with many layers by using **backpropagation**.

Convolutional networks may include local or global pooling layers, which combine the outputs of neuron clusters at one layer into a single neuron in the next layer. For example, max pooling uses the maximum value from each of a cluster of neurons at the prior layer. Another example is average pooling, which uses the average value from each of a cluster of neurons at the prior layer.

Fully connected layers connect every neuron in one layer to every neuron in another layer. It is in principle the same as the traditional Multi-Layer Perceptron neural network (MLP). The **flattened matrix** goes through a fully connected layer to classify the images.

In neural networks, each neuron receives input from some number of locations in the previous layer. In a fully connected layer, each neuron receives input from every element of the previous layer. In a convolutional layer, neurons receive input from only a restricted **subarea** of the previous layer. Typically the subarea is of a square shape (e.g., size 5 by 5). The input area of a neuron is called its receptive field. So, in a fully connected layer, the receptive field is the entire previous layer. In a convolutional layer, the receptive area is smaller than the entire previous layer.

Each neuron in a neural network computes an output value by applying some function to the input values coming from the receptive field in the previous layer. The function that is applied to the input values is specified by a vector of weights and a **bias** (typically real numbers). Learning in a neural network progresses by making incremental adjustments to the biases and weights. The vector of weights and the bias are called a filter and represents some feature of the input (e.g., a particular shape). A distinguishing feature of CNNs is that many neurons share the same filter. This reduces **memory footprint** because a single bias and a single vector of weights is used across all receptive fields sharing that filter, rather than each receptive field having its own bias and vector of weights.

 **Words**

```
imagery ['imidʒəri] n. 影像,意象,形象化        tile [tail] v. 平铺显示
shared-weight    参数共享                      gradient ['greidiənt] n. 梯度,渐变
cortex ['kɔːteks] n. 皮质,皮层                backpropagation    反向传播
cortical ['kɔːtikəl] adj. 皮层的,皮质的        subarea ['sʌbɛəriə] n. 分区
stimuli ['stimjulai] n. 刺激,刺激物           bias ['baiəs] n. 偏差
hand-engineered    人工提取的
```

 **Phrases**

```
multilayer perception    多层感知器
shift invariant    移位不变性
visual field    视野
receptive field    感受野
recommender system    推荐系统
pooling layer    池化层
by convention    按照惯例
flattened matrix    扁平矩阵
memory footprint    内存占用
```

 **Notes**

[1] 线性整流函数(Rectified Linear Unit, ReLU),又称修正线性单元,是一种人工神经网络中常用的激活函数(activation function),通常指代以斜坡函数及其变种为代表的非线性函数。

## Exercises

Ⅰ. **Read the following statements carefully, and decide whether they are true (T) or false (F) according to the text.**

____ 1. In a convolutional layer, the receptive area is the entire previous layer.

____ 2. In a fully connected layer, the receptive field is smaller than the entire previous layer.

____ 3. Each convolutional neuron processes data only for its visual field.

____ 4. Neurons receive input from only a restricted subarea of the previous layer in a convolutional layer.

____ 5. Each neuron receives input from some number of locations in the previous layer in neural networks.

Ⅱ. **Choose the best answer to each of the following questions according to the text.**

1. Which of the following is right regarding the neural network?（    ）
   A. In a convolutional layer, the receptive area is the entire previous layer.
   B. In a fully connected layer, the receptive field is smaller than the entire previous layer.
   C. Each neuron receives input from some number of locations in the previous layer.
   D. All of the above

2. What kinds of layers does a convolutional neural network include?（    ）
   A. An input layer
   B. An output layer
   C. Multiple hidden layers
   D. All of the above

3. How many weights does a fully connected layer for a (small) image of size 100 × 100 have for each neuron in the second layer?（    ）
   A. 100
   B. 1000
   C. 10000
   D. None of the above

Ⅲ. **Fill in the numbered spaces with the words or phrases chosen from the box. Change the forms where necessary.**

> only   big   recurrent   beyond   general   has
> object   other   approach   rather

**Graph Neural Network**

A graph neural network is the "blending powerful deep learning ____1____ with structured representation" models of collections of objects, or entities, whose relationships are explicitly mapped out as "edges" connecting the objects. Human cognition makes the strong assumption that the world is composed of ___2___ and relations, and because GNs (Graph Networks) make a similar assumption, their behavior tends to be more interpretable. ___3___, modeling the relationships of objects is something that not ___4___ spans all the various machine learning models — CNNs, Recurrent Neural Networks (RNNs), Long-Short-Term Memory (LSTM) systems, etc.— but also ___5___ approaches that are not neural nets, such as set theory. The idea is that graph networks are ___6___ than any one machine-learning approach. Graphs bring an ability to generalize about structure that the individual neural nets don't ___7___. Graphs, ___8___, are a representation which supports arbitrary (pairwise) relational structure and computations over graphs afford a strong relational inductive bias ___9___ that which convolutional and ___10___ layers can provide.

IV. Translate the following passage into Chinese.

**Regularization**

Regularization is a way to avoid overfitting by penalizing high-valued regression coefficients. In simple terms, it reduces parameters and shrinks (simplifies) the model. This more streamlined, more parsimonious model will likely perform better at predictions. Regularization adds penalties to more complex models and then sorts potential models from least overfit to greatest; The model with the lowest "overfitting" score is usually the best choice for predictive power.

## Part 2

# Simulated Writing: Writing Professional Letters (I)

商务信函是一种有效的方式,被用来传递正式的或有说服力的信息。它能够建立永久性的记录,或者能够发送重要、敏感或机密的信息。尽管电子邮件已经成为最流行的交换书面信息的方式,但商务信函仍然是一个必要的通信工具。通常写信是为了与所处的组织之外的人进行沟通,当然,也可以通过写信给同事发送正式的消息。除了页面上所写的话,信的设计和格式也在向读者展示作者自己,作者对细节的关注,以及作者的专业水平。

1. 如何撰写商务信函

商务信函是一种在组织外传递信息的专业交流工具。虽然商务信函的使用频率远低于其他的交流媒介,如电子邮件和传真,但当需要与供应商、其他商家,最重要的是与客户交流时,商务信函是最合适的选择。图7-3显示了典型的商务信函的例子。

在以下情况,可以写一封商务信函:

(1) 与不认识的人进行交流。如果需要和一个没见过或根本不认识的人进行交流,那么最好发一封商务信函来建立职业关系。虽然电子邮件很容易撰写,而且送达速度更快,但

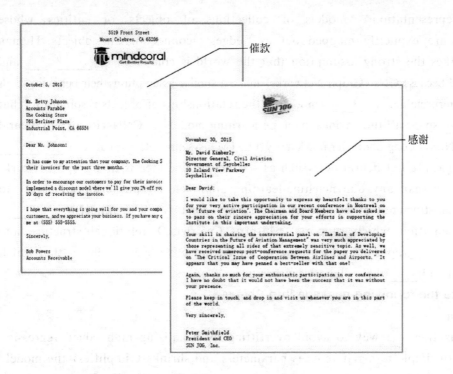

图 7-3 商务信函的例子

其非正式性及自发性会使邮件显得太自我或太大胆。

（2）记录交流。如果需要保持与公司以外的人进行正式交流的书面记录，那么通常商务信函是最好的选择。商务信函可以提供一个永久的记录，尤其是在附有合同、协议条款，或特别优惠时。

(3) 传达坏消息或讨论敏感的事情。打印在公司信纸上的商务信函会比其他渠道如电子邮件更正式,也更能表示尊重。撰写书面的信能够向读者展示作者对这个问题的认真态度。此外,商务信函可以保密,并且比一些数字通信的形式拥有更多的隐私。

(4) 表达善意。当想要表示感谢、祝贺、慰问,或道歉时,书面信是十分合适的。在每一种情况下,一封信,包括信纸、字体和签名,都会比一封非正式的信更加有效地表达作者的情感。

在以下情况,可以打电话、亲自访问或发送电子邮件:

(1) 尽快发出信息。商务信函通常通过最好的邮递工具进行发送,这可能需要几天才能送达。隔夜快递服务是一个选项,虽然递送成本很高。

(2) 与有良好工作关系的人联系。用信件与认识的人进行日常交流有些过于正式,但当想表达善意或需要书面记录时就是例外了。

(3) 写一个常规的主题。电子邮件的流行是因为它的高效,而打电话和直接访问会比书面的信息更加个性化。对于常规的交流,如对于不需要创建永久性记录、保密、表示礼节,或提供有说服力的论据的请求和响应,请使用电子邮件或打电话。

**2. 书写商业信函**

在写一封商务信函之前,首先要确立沟通的目标或目的——是正在发出请求,响应查询,记录决策,还是确认一个行动？然后,可以考虑并预期读者对所写的信函的反应。当开始撰写的时候,可以遵循图 7-4 所示的商务信函的标准格式。作者应使用正确的格式,或者用齐头式来展示其专业性,并且简化任务。

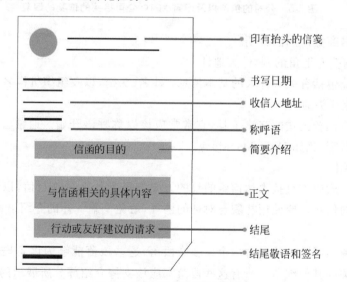

图 7-4　标准的商务信函格式

1) 印有抬头的信笺

大多数商务信函都是写在印有抬头的信笺纸上,上面含有公司的名称、地址、电话和传真号码,以及网址。通常还会有企业徽标来标志该组织。图 7-5 展示了 XETC 公司的信笺。

2) 书写日期

以今天的日期开头,再写上月份,并使用四位数表示年份。信最好使用当前的日期。不

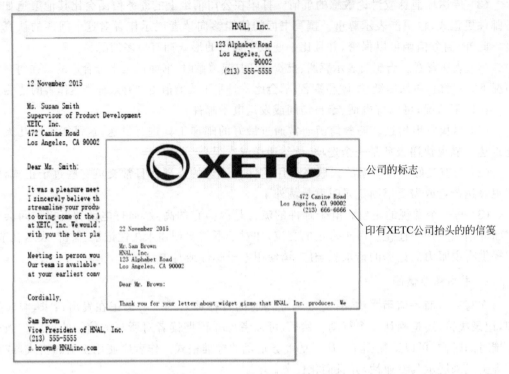

图 7-5　公司的信函以及印有 XETC 公司抬头的信笺的回复

要提早或者推迟商务信函的日期。

3)（写在信笺左上角的）收件人地址

收件人地址包括有关收件人的基本信息：姓名、头衔，以及组织机构名称和邮寄地址。

4)（书信等开头的）称呼语

商务信函是比较正式的交流工具，它常常应该以称呼语开始。通常会说 Dear，后面紧跟着收件人的名字，比如 Dear Ms. Mary。

5) 简要介绍

信函的第一段应该直接表达信函的目的。要解释为什么写这封信，以便收件人可以更好地了解信中的信息。要使用礼貌与对话的语气，避免千篇一律的介绍语言。

6) 正文

大多数商务信函都应该包含一个或多个段落，这些段落能够为读者提供信息、解释，或者与此信函相关的其他细节。所有这些段落都应该支持介绍部分所展示的主旨。

7) 结尾

还需要用结尾段优雅地结束这封信，切忌突兀地结束商务信函。相反，以善意的表达、礼貌的评论或意见，或要求采取具体行动的请求来结尾会赢得他人的好感。

8) 结尾敬语和签名

在信的结尾用敬语，如真诚地（Sincerely）、恭敬地（Respectfully）或亲切地（Cordially）。之后在结尾敬语下面的四行插入名字，这四行为手写签名留够了空间。

3. 称呼语的使用

当撰写一封商务信函时,也是在建立一个你自己的以及你所代表的组织的形象。通常,信函决定了第一次和你接触的人对你的持久的印象。以用适当的称呼语和介绍开始商务信函可以建立这封信件友好的基调,而且可以使读者留下深刻的印象,如图7-6所示。

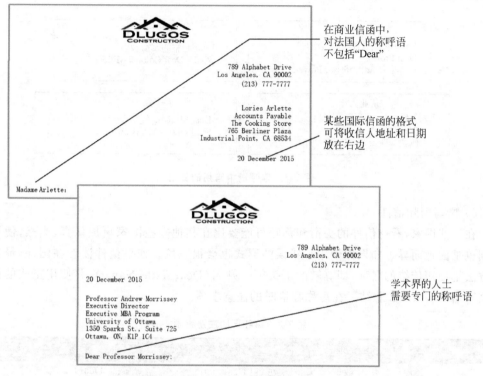

图 7-6　称呼语中常用的头衔

1) 称呼格式

尊敬的＋头衔＋姓名(Dear Title Name)的称呼格式在任何商务信函中都是正确的,应该用在所有的商务信函中。使用先生(Mr.)、女士(Ms.)的尊称,除非收件人是一个有着正式的头衔的人,如博士(Dr.)或牧师(Reverend)。

2) 标点符号

称呼语中的标点也代表着信的意图。商务信函在称呼语之后写一个冒号(：),而私人信件则用逗号(,)。

3) 名字

写一封正式的信件时,称呼中不能使用收件人的名字,如尊敬的路易莎·琼斯女士(Dear Ms. Louisa Jones)或尊敬的卡尔·罗伯茨先生(Dear Mr. Carl Roberts)。然而,如果你和收件人关系比较友好,则可以使用名字,如尊敬的鲍勃(Dear Bob)。

4) 非个人称呼

如果你并不知道收件人的名字,那就使用一个非个人称呼,如读者的头衔,如尊敬的运营经理(Dear Operations Manager)的名称或使用他们的部门或单位的名称,如尊敬的人力资源部(Dear Human Resources Department)。

5) 等级和荣誉头衔

在撰写特别正式的商务信函时,你可能需要在称呼中加入工作头衔、等级或荣誉称号,如尊敬的坎宁安总统(Dear President Cunningham)、尊敬的史密斯博士(Dear Dr. Smith)或者尊敬的华顿大使(Dear Ambassador Wharton)。送到政要处的信件的称呼中可以加入尊敬的(Honorable)或阁下(Excellency)。图7-7中列出了一些常用的称呼语。

```
┌─────────────────────────────────────┬─────────────────────────────────────┐
│ 学术界                              │ 陆海空三军                          │
│ Professor 名姓(收信人地址)          │ Dear 完整的军衔 姓:                 │
│ Dear Professor 姓                   │ 例如:Dear Colonel Miller:          │
│ 例如:Dear Professor Powers         │                                     │
├─────────────────────────────────────┼─────────────────────────────────────┤
│ 从业人士                            │ 社会                                │
│ 名姓 M.D.(收信人地址)               │ Dear Mr./Ms. 姓                     │
│ Dear Dr. 姓                         │ 例如:Dear Ms. Brown:               │
│ 例如:Dear Dr. Smith:               │                                     │
└─────────────────────────────────────┴─────────────────────────────────────┘
```

图 7-7 称呼语中常用的头衔

6) 撰写国际信件

在一些国家,看待信件的头衔和称呼可能要比在其他一些国家更加认真,当然,使用的规则也是因地而异。如果你使用了错误的形式也会很尴尬。如果信件读者在国外,最安全的方法是使用传统的名称,即尊敬的 + 头衔 + 姓名(Dear Title Name),并使用正式的语气去撰写信函。表7-2总结了在写称呼语时的注意事项。

表 7-2 称呼语的注意事项

| 称 呼 语 | 适合提到 | 尽量避免 |
| --- | --- | --- |
| 格式 | • 使用尊敬的 + 头衔 + 姓名(Dear Title Name)的称呼格式<br>• 在商务信函的称呼中接一个冒号(:) | • 省略尊敬的(Dear)<br>• 在结尾处使用逗号,除了私人信函之外 |
| 姓名 | • 使用收件人的姓<br>• 仅使用收件人的名字(用于私人信函) | • 用女士(Miss)或夫人(Mrs.)作为收件人头衔<br>• 在正式信函中使用收件人的名字 |
| 国际收件人 | • 熟悉收件人国家的写信习俗<br>• 使用正式的称呼语 | • 信函以个人为中心<br>• 你与收件人有着友好的关系,使用非正式的称呼语 |

转 147 页

# Part 3

# Listening & Speaking

## Dialogue:Deep Learning

(As deep learning is becoming more and more popular,Henry would like to know

Unit 7　Deep Learning

*more about it.)*

**Henry:** Excuse me, Sophie and Mark. Could you tell me about deep learning? It is a **buzz** word nowadays.

**Sophie:** Sure. Deep learning is part of a broader family of machine learning methods based on learning data representations, as opposed to [1] task-specific algorithms. Learning can be supervised, semi-supervised or unsupervised.

[1] Replace with:
1. as against
2. set against
3. against

**Mark:** To my understanding,[2] deep learning architectures such as deep neural networks, deep belief networks and recurrent neural networks have been applied to fields including computer vision, speech recognition, natural language processing, audio recognition, social network filtering, machine translation, **bioinformatics**, drug design, medical image analysis, material inspection and **board game** programs, where they have produced results comparable to and in some cases superior to human experts.

[2] Replace with:
1. To what I know,
2. Based on my understanding,

**Sophie:** And deep learning models are vaguely inspired by information processing and communication patterns in biological nervous systems yet have various differences from the structural and functional properties of biological brains (especially human brains), which make them incompatible with neuroscience evidences.

**Henry:** So what are the functions of deep learning?

**Mark:** Well, deep learning is a class of machine learning algorithms that use **a cascade of** multiple layers of nonlinear processing units for feature extraction and transformation. Each successive layer uses the output from the previous layer as input.

**Sophie:** And these machine learning algorithms learn in supervised (e.g., classification) and/or unsupervised (e.g., pattern analysis) manners. And also they learn multiple levels of representations that correspond to different levels of abstraction; the levels form a hierarchy of concepts.

Henry: Could you please describe it in more detail, Sophie?

Sophie: Ok. In deep learning, each level learns to transform its input data into a slightly more abstract and composite representation. In an image recognition application, the raw input may be a matrix of pixels; the first representational layer may **abstract** the pixels and encode edges; the second layer may compose and encode arrangements of edges; the third layer may encode a nose and eyes; and the fourth layer may recognize that the image contains a face. Importantly, a deep learning process can learn which features to optimally place in which level on its own.

Mark: Of course, this does not completely **obviate** the need for **hand-tuning**; for example, varying numbers of layers and layer sizes can provide different degrees of abstraction.

Henry: And what does the "deep" in "deep learning" mean?

Sophie: Well, the "deep" in "deep learning" refers to the number of layers through which the data is transformed. More precisely, deep learning systems have a substantial **Credit Assignment Path** (CAP) depth.

Henry: A CAP?

Sophie: Yes, The CAP is the chain of transformations from input to output. CAPs describe potentially causal connections between input and output. For a feedforward neural network, the depth of the CAPs is that of the network and is the number of hidden layers plus one (as the output layer is also **parameterized**).

Mark: And for recurrent neural networks, in which a signal may propagate through a layer more than once, the CAP depth is potentially unlimited. No universally agreed upon threshold of depth divides shallow learning from deep learning, but most researchers agree that deep learning involves CAP depth>2. CAP of depth 2 has been shown to be a

|  | **universal approximator** in the sense that it can emulate any function. |  |
|---|---|---|
| Sophie: | Beyond that more layers do not add to the function approximator ability of the network. Deep models (CAP>2) are able to extract better features than shallow models and hence,[3] extra layers help in learning features. | [3] Replace with:<br>1. therefore,<br>2. thus,<br>3. for this reason, |
| Henry: | I've got it. Thanks for taking your time to explain to me. |  |
| Sophie & Mark: | It's OK. |  |

## Exercises

Work in a group, and make up a similar conversation by replacing the statements with other expressions on the right side.

## Words

buzz[bʌz] n. 时髦的（词语、想法或活动）
bioinformatics[ˌbaɪəʊɪnfəˈmætɪks] n. 生物信息学
abstract[ˈæbstrækt] v. 提取

obviate[ˈɒbvieɪt] v. 排除，避免
hand-tuning 手动调谐
parameterized[pəˈræmɪtəraɪzd] adj. 参数化的

## Phrases

board game    棋盘游戏
a cascade of    一系列，一连串
credit assignment path    权重分配路径
universal approximator    泛逼近器

## Listening Comprehension: Generative Adversarial Network

在线音频

Listen to the article and answer the following 3 questions based on it. After you hear a question, there will be a break of 15 seconds. During the break, you will decide which one is the best answer among the four choices marked (A),(B),(C) and (D).

**Questions**

1. How many neural networks does a generative adversarial network include? (     )
   (A) One

(B) Two
(C) Three
(D) Four

2. Which of the following is the abbreviation of GAN according to this text?（　　）

(A) Global Advanced Network
(B) General Analysis Network
(C) Generative Adversarial Network
(D) None of the above

3. Which of the following does not belong to generative adversarial network?（　　）

(A) A convolutional neural network
(B) A generative network
(C) A discriminative network
(D) All of the above

## Words

discriminative[disˈkrimɪnətiv] *adj*. 区分的，区别的

authentic[ɔːˈθentik] *adj*. 真正的，真实的，可信的

stubborn[ˈstʌbən] *adj*. 难处理的，顽固的

## Abbreviations

GAN　Generative Adversarial Network　生成对抗网络

## Dictation：Recurrent Neural Network

*This article will be played three times. Listen carefully, and fill in the numbered spaces with the appropriate words you have heard.*

A Recurrent Neural Network（RNN）is a class of artificial ___1___ network where connections between nodes form a ___2___ graph along a sequence. This allows it to ___3___ **temporal** dynamic behavior for a time ___4___ . Unlike feedforward neural networks, RNNs can use their internal state (memory) to ___5___ sequences of inputs. This makes them ___6___ to tasks such as unsegmented, connected handwriting ___7___ or speech recognition.

The term "recurrent neural network" is used **indiscriminately** to ___8___ to two broad classes of networks with a similar ___9___ structure, where one is finite impulse and the other is ___10___ impulse. Both classes of networks exhibit temporal dynamic

11    . A finite impulse recurrent network is a directed **acyclic** graph that can be **unrolled** and replaced with a    12    feedforward neural network, while an infinite impulse recurrent network is a directed    13    graph that can't be unrolled.

Both finite impulse and infinite impulse recurrent networks can have additional    14    state, and the storage can be    15    direct control by the neural network. The storage can also be    16    by another network or graph, if that    17    time delays or has feedback    18   . Such controlled states are referred to as    19    state or gated memory, and are part of Long    20    Memory Networks (LSTMs) and **gated recurrent units**.

## ▶ Words

temporal ['tempərəl] *adj.* 暂时的,时间的
indiscriminately [ˌindi'skriminətli] *adv.*
　不加选择地,任意地

acyclic [ˌei'saiklik] *adj.* 非循环的
unroll [ʌn'rəul] *v.* 展开,显示

## ▶ Phrases

gated recurrent unit　门控循环单元

# Unit 8

## Reinforcement Learning

Unit 8　Reinforcement Learning

# Reading & Translating

## Section A：Reinforcement Learning, Deep Learning's Partner

These years, we have seen all the **hype** around AI Deep Learning. With recent innovations, deep learning demonstrated its usefulness in performing tasks such as image recognition, voice recognition, price forecasting, across many industries. It's easy to overestimate deep learning's capabilities and pretend it's the magic bullet that will allow AI to obtain General Intelligence. In truth, we are still far away from that. However, deep learning has a relatively unknown partner: Reinforcement Learning (Figure 8-1). As AI researchers **venture** into the areas of Meta-Learning [1], attempting to give AI learning capabilities, **in conjunction with** deep learning, reinforcement learning will play a crucial role.

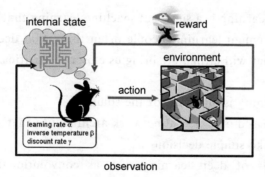

Figure 8-1　Reinforcement Learning

Imagine a child who is learning by interacting with their environment. Each touch will generate a sensation that can result in a reward. For instance, the pleasant smell of the flower will **entice** the child to want to smell the flower again; the pain from a **prick** of the flower's **stem** will alert the child who will **refrain from** touching the stem again.

In each case, as the child interacts with the environment, the environment **reciprocates** and teaches the child by rewarding the child with different sensations.

The child is learning by trial and error.

This is reinforcement learning. In reinforcement learning, an agent starts in a neutral state. Then, as actions are taken, the environment helps the agent transition from the neutral state to other states. In these other states, there might be rewards for the agent.

The goal of the agent is to gather as many rewards as possible.

You can visualize yourself as an agent, walking on a reinforcement learning path, starting at the beginning of a maze toward the exit. With each step that you take, you have a chance of collecting rewards that you can **tally up**. Depending on the type of rewards and the quantity of the rewards, in your reward **pouch**, decisions can be made to direct you toward the exit. Eventually, with many tries, an optimal path can be found through the maze.

Reinforcement learning is already widely used in many industries such as manufacturing, inventory management, delivery management, and finance.

In manufacturing, robots use reinforcement learning to learn specific tasks in sequence with precision on the assembly line.

In delivery management systems, reinforcement learning can be used to split the customer's order among different vehicles on different routes to arrive at the destination.

In the financial industry, reinforcement learning is used to evaluate trading strategies to fulfill financial objectives.

Most of the time, these reinforcement learning algorithms are integrated with deep learning algorithms to create deep reinforcement learning algorithms that can handle more complex tasks.

In contrast, deep learning is a subset of machine learning that's receiving a lot more attention than reinforcement learning. People often think that deep learning is the part of machine learning that will ultimately bring us closer to Artificial General Intelligence (AGI).

In truth, deep learning is only part of the solution.

Deep learning is based on a neural network architecture that's similar to the human brain to allow it to make simple decisions.

Traditional models of deep learning such as convolutional neural networks or recurrent neural networks are great at making simple decisions and finding hidden relationships. However, due to the restrictions in the number of layers of complexity, time, and data, deep learning is limited in its capacity to make decisions in complex situations.

On top of that, deep learning by itself cannot infer the meaning of the patterns that it sees. Most technologists will agree that deep learning, by itself is great for classification problems, but it is inadequate for problems that require reasoning, understanding and common sense.

Deep learning also has a huge limitation: it needs to consume a lot of data to be able to make accurate decisions. The biggest asset in human decision making is that our brains can **zoom in on** "critical" information. Often, we make important decisions on little current information in conjunction with our past experiences and knowledge from the past.

Artificial General Intelligence can only become more sophisticated if it can also make decisions based on small amounts of data.

In other words, deep learning by itself does not bring us to Artificial General Intelligence. It needs a partner that can perform abstractions and reasoning.

One of the most important aspects of research into Artificial General Intelligence is in the area of reinforcement learning. While Deep Learning gives AGI the ability to uncover hidden patterns to make connections, it is reinforcement learning that allows AGI to make abstractions to understand the meaning behind patterns and **in turn** direct behavior.

When deep learning is combined with reinforcement learning in deep reinforcement learning, the AGI can plan, understand, and **strategize** the actions that it should take.

An example of this is DeepMind's MuZero algorithm, a deep reinforcement learning algorithm that's able to construct agents that can **plan out** how to play games such as chess and **GO**, without knowing the rules.

This is the first step of artificial general intelligence.

When AI can use limited data to plan, understand and strategize actions, without being explicitly "taught", then the AI is closer to achieving general intelligence.

In the coming year, we will likely see a lot more applications of deep reinforcement learning algorithms in different industries. This is an exciting time. With many applications in different industries, the usage of these algorithms will become more prevalent and sophisticated. With many iterations of research and application, we can truly see the potential of power of AI that might someday achieve general intelligence.

## Words

hype[haip] *n.* 大肆宣传,炒作
venture['ventʃə(r)] *v.* 敢于冒险
entice[in'tais] *v.* 引诱,诱惑,吸引
prick[prik] *n.* 刺
stem[stem] *n.* 干,茎

reciprocate[ri'siprəkeit] *v.* 互换,报答
pouch[pautʃ] *n.* 小袋
strategize['strætidʒaiz] *v.* 制定战略
go[gəu] *n.* 围棋

## Phrases

in conjunction with     连同,与……协力
refrain from     忍住,抑制,制止
tally up     结算
zoom in on     聚焦于,推近

in turn    依次，反过来
plan out   策划，为……做准备

### Notes

[1] 和之前的机器学习算法相比，元学习（meta-learning）是让机器学会学习。传统的机器算法只是希望机器学习到一个函数 $f$，通过这个函数 $f$，来判断这个图片究竟属于哪个类别，如一条微博的评论是正面的还是负面的。但是元学习是希望机器在学习了一些图片分类任务之后，可以学会文本分类的任务。

### Exercises

I. Read the following statements carefully, and decide whether they are true (T) or false (F) according to the text.

____ 1. Convolutional neural networks or recurrent neural networks are traditional models of deep learning.

____ 2. Deep learning by itself brings us to Artificial General Intelligence.

____ 3. Reinforcement learning allows AGI to make abstractions to understand the meaning behind patterns and in turn direct behavior.

____ 4. Reinforcement learning is one of the most important aspects of research into Artificial General Intelligence.

____ 5. Deep learning is based on a neural network architecture to allow it to make simple decisions.

II. Choose the best answer to each of the following questions according to the text.

1. Which of the following is deep learning based on?（     ）
   A. A neural network
   B. A Wi-Fi Network
   C. A local area network
   D. A wide area network

2. Which of the following is a deep reinforcement learning algorithm?（     ）
   A. Decision tree
   B. DeepMind's MuZero
   C. SVM
   D. K-means

3. Which of the following is not right?（     ）
   A. Deep learning is a subset of machine learning.
   B. Machine learning is a subset of deep learning.
   C. Deep learning, by itself is great for classification problems.
   D. Deep learning is inadequate for problems that require reasoning, understanding and common sense.

## Unit 8  Reinforcement Learning

**III. Fill in the numbered spaces with the words or phrases chosen from the box. Change the forms where necessary.**

> use   learn   advance   strategy   convolution
> type   model   involve   find   order

**Q-learning**

Q-learning is a term for an algorithm structure representing model-free reinforcement learning. By evaluating policy and using stochastic modeling, Q-learning ___1___ the best path forward in a Markov decision process.

The technical makeup of the Q-learning algorithm ___2___ an agent, a set of states and a set of actions per state.

The Q function ___3___ weights for various steps in conjunction with a discount factor in ___4___ to value rewards.

Although it may seem like a simple idea, Q-learning is of paramount importance in many ___5___ of reinforcement learning and deep learning models. One of the best examples is where deep Q-learning is used to help machine ___6___ programs to learn game-play ___7___ in various types of video games, for example, in Atari games from the 1980s. Here a ___8___ neural network takes samples of game-play in order to work up a stochastic ___9___ that will help the computer know how to play the game better over time.

Q-learning has abundant potential for helping to ___10___ artificial intelligence and machine learning.

**IV. Translate the following passage into Chinese.**

**Markov Decision Process**

Reinforcement Learning is a type of Machine Learning. It allows machines and software agents to automatically determine the ideal behavior within a specific context, in order to maximize its performance. Simple reward feedback is required for the agent to learn its behavior; this is known as the reinforcement signal.

There are many different algorithms that tackle this issue. As a matter of fact, Reinforcement Learning is defined by a specific type of problem, and all its solutions are classed as Reinforcement Learning algorithms. In the problem, an agent is supposed to decide the best action to select based on his current state. When this step is repeated, the problem is known as a Markov Decision Process.

## Section B: AlphaGo Zero: Starting from Scratch

It is able to do this by using a **novel** form of reinforcement learning, in which AlphaGo Zero becomes its own teacher. The system **starts off with** a neural network that knows nothing about the game of Go. It then plays games against itself, by combining this neural network with a powerful search algorithm. As it plays, the neural network is **tuned** and updated to predict moves, as well as the eventual winner of the games.

This updated neural network is then recombined with the search algorithm to create a new, stronger version of AlphaGo Zero, and the process begins again. In each iteration, the performance of the system improves by a small amount, and the quality of the self-play games increases, leading to more and more accurate neural networks and ever stronger versions of AlphaGo Zero.

This technique is more powerful than previous versions of AlphaGo because it is no longer constrained by the limits of human knowledge. Instead, it is able to learn **tabula rasa** from the strongest player in the world: AlphaGo itself.

It also differs from previous versions in other **notable** ways.

• AlphaGo Zero only uses the black and white stones from the Go board as its input, whereas previous versions of AlphaGo included a small number of hand-engineered features.

• It uses one neural network rather than two. Earlier versions of AlphaGo used a "policy network" to select the next move to play and a "value network" to predict the winner of the game from each position. These are combined in AlphaGo Zero, allowing it to be trained and evaluated more efficiently.

• AlphaGo Zero does not use "**rollouts**"—fast, random games used by other Go programs to predict which player will win from the current board position. Instead, it relies on its high quality neural networks to evaluate positions.

All of these differences help improve the performance of the system and make it more general. But it is the algorithmic change that makes the system much more powerful and efficient.

AlphaGo has become progressively more efficient thanks to hardware gains and more recently algorithmic advances.

After just three days of self-play training, AlphaGo Zero **emphatically** defeated the previously published version of AlphaGo—which had itself defeated 18-time world champion Lee Sedol—by 100 games to 0. After 40 days of self-training, AlphaGo Zero became even stronger, outperforming the version of AlphaGo known as "Master", which has defeated the world's best players and world number one Ke Jie.

ELO Ratings [1]—A measure of the relative skill levels of players in competitive games such as go—show how alphago has become progressively stronger during its development.

Over the course of millions of AlphaGo vs AlphaGo games, the system progressively learned the game of Go **from scratch** (Figure 8-2), accumulating thousands of years of human knowledge during a period of just a few days. AlphaGo Zero also discovered new knowledge, developing unconventional strategies and creative new moves that **echoed** and surpassed the novel techniques it played in the games against Lee Sedol and Ke Jie.

These moments of creativity give us confidence that AI will be a **multiplier** for human **ingenuity**, helping us with our mission to solve some of the most important

Figure 8-2　AlphaGo Zero：Starting from scratch

challenges humanity is facing.

　　While it is still early days, AlphaGo Zero constitutes a critical step towards this goal. If similar techniques can be applied to other structured problems, such as protein folding, reducing energy consumption or searching for revolutionary new materials, the resulting breakthroughs have the potential to positively impact society.

## Words

| | |
|---|---|
| novel['nɒvl] *adj*. 新奇的,异常的 | 然地 |
| tune[tjuːn] *v*. 调整,使一致 | echo['ekəʊ] *v*. 重复,附和(想法或看法) |
| notable['nəʊtəbl] *adj*. 值得注意的,显著的 | multiplier['mʌltɪplaɪə(r)] *n*. 乘数 |
| rollout 首次展示 | ingenuity[ˌɪndʒə'njuːəti] *n*. 独创性,足智多谋 |
| emphatically[ɪm'fætɪkli] *adv*. 着重地,断 | |

## Phrases

from scratch　　白手起家,从头做起
start off with　　从……开始,用……开始
tabula rasa　　白板(没有写字的书写板)

## Notes

　　[1] 埃洛等级分系统(ELO Rating System)是当今对弈水平评估的公认权威规则,已被广泛应用于国际象棋、围棋、足球和篮球等体育运动以及游戏中。例如,星际争霸天梯排行、魔兽世界竞技场、Dota 天梯系统、LOL 匹配等竞技比赛系统中。ELO 是一套较为完善的评分规则和机制,比较适合对竞技类游戏的选手的技术等级进行评估,用以计量个体在对决类比赛中相对技能的算法系统,对于游戏而言,需要让每场游戏尽可能地接近公平,创造双方势均力敌的竞赛环境。它最初由美国物理学教授 Arpad Elo 创立,故命名为埃洛排名。

# Exercises

Ⅰ. Read the following statements carefully, and decide whether they are true (T) or false (F) according to the text.

  ____ 1. AlphaGo Zero relies on its high quality neural networks to evaluate positions.

  ____ 2. As AlphaGo Zero plays, the neural network is tuned and updated to predict moves, as well as the eventual winner of the games.

  ____ 3. AlphaGo Zero includes a small number of hand-engineered features.

  ____ 4. AlphaGo only uses the black and white stones from the Go board as its input.

  ____ 5. AlphaGo Zero has defeated the world's best players and world number one Ke Jie,

Ⅱ. Choose the best answer to each of the following questions according to the text.

1. Which of the following has defeated the world's best players and world number one Ke Jie? (　　)

  A. AlphaGo

  B. AlphaGo Zero

  C. ELO Rating

  D. None of the above

2. How many neural networks does AlphaGo Zero use? (　　)

  A. One

  B. Two

  C. Three

  D. Four

3. Which of the following is not right? (　　)

  A. AlphaGo Zero relies on its high quality neural networks to evaluate positions.

  B. As AlphaGo Zero plays, the neural network is tuned and updated to predict moves, as well as the eventual winner of the games.

  C. AlphaGo Zero has defeated the world's best players and world number one Ke Jie,

  D. AlphaGo Zero includes a small number of hand-engineered features.

Ⅲ. Fill in the numbered spaces with the words or phrases chosen from the box. Change the forms where necessary.

| implicit take agent short base |
| --- |
| train deal correct able negative |

# Unit 8  Reinforcement Learning

**Reinforcement Learning for Chatbots**

Typically Chatbots are ___1___ with the data to answer certain questions. There are obviously outliers that can possibly come when a chatbot is not ___2___ to find relevant intent for the given utterance from the user. A Reinforcement Learning ___3___ can be possibly set up to give a guideline to the answer and the possible outcome could be positive or ___4___. The agent can add to the training data. There are two reasons to use Reinforcement Learning in chatbot. Chatbots usually has an ___5___ or explicit feedback mechanism to say whether the answer is correct or not ___6___. The second major reason is the time ___7___ to train on a huge knowledge base that exists in an enterprise. Most of the enterprises face the challenge of how to train the chatbot with their existing knowledge ___8___.

A chatbot designed with the Q-learning model to ___9___ with the outliers will help to generate more training data to learn. In ___10___ Chatbot can autonomously learn over a period of time using the Q-learning models.

**IV. Translate the following passage into Chinese.**
**Why "Deep" Q-Learning?**

Q-learning is a simple yet quite powerful algorithm to create a cheat sheet for our agent. This helps the agent figure out exactly which action to perform.

But what if this cheat sheet is too long? Imagine an environment with 10 000 states and 1000 actions per state. This would create a table of 10 million cells. Things will quickly get out of control!

It is pretty clear that we can't infer the Q-value of new states from already explored states. This presents two problems:

- First, the amount of memory required to save and update that table would increase as the number of states increases
- Second, the amount of time required to explore each state to create the required Q-table would be unrealistic

Here's a thought—what if we approximate these Q-values with machine learning models such as a neural network? Well, this was the idea behind DeepMind's algorithm that led to its acquisition by Google for 500 million dollars!

## Part 2

# Simulated Writing: Writing Professional Letters (II)

接 132 页

**4. 结束商业信函**

结束商务信函的方式会影响读者对信件的理解、对请求的意愿以及他们对你的印象。商务信函的最后总是以谦称结尾，如真诚地(Sincerely)。签名档应该出现在商务信函的结尾，由签名、打印的名字和头衔（如果你正在写一封正式信函）组成。图 8-3 即为正式的商务信函的结尾例子。

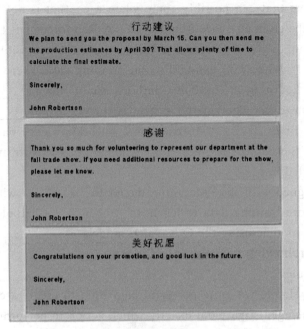

图 8-3 结尾的例子

1）结束时呼吁行动

读者通常会浏览信函的最后一段以寻找行动要求、期限和你要求他们完成的工作。截止日期特别有用。记得在要求时要有礼貌，因为人们在受到尊重时，反应会更为亲切，表 8-1 总结了商务信函的结尾和签名档的注意事项。

表 8-1 结尾和签名档的注意事项

| 结尾元素 | 适合提到 | 尽量避免 |
| --- | --- | --- |
| 结尾段 | • 在提出请求时要具体而有礼貌<br>• 包括截止日期<br>• 为请求和截止日期提供理由<br>• 通过提供联系方式使回复更容易 | • 命令读者尽快回复<br>• 以"感谢您关注此事"(Thank you for your attention to this matter)这样老套的语言结尾 |
| 谦称结尾 | • 大部分商务信函使用传统的真诚地(Sincerely)结尾<br>• 一些个人信函可以使用替换词亲切地(Cordially)<br>• 在跨国信函中使用恭敬地(Respectfully)来表达敬意 | • 以消极的情绪结尾，比如愤怒地(Angrily)或失望地(Disappointedly)<br>• 省略正式信件之后的结尾和逗号 |
| 签名档 | • 在正式信函中，手写并打印你的全名<br>• 如果收件人地址写明了收件人的头衔，那么你也要写清你的头衔<br>• 如果代表公司，那么写上公司的全名 | • 在正式信件中，只写你的名字<br>• 签缩写名字<br>• 使用计算机合成的名字 |

2）表示感谢

在要求某事的同时要表示感谢。可以在请求中直接感谢,如"若能在 6 月 15 日前完成这个报告,我将不胜感激。"(I appreciate your help in completing this report by June 15.)

3）包含善意

如果不做具体要求,则以积极的陈述、观察,或追求一个持续的关系来结尾。即使写的是一个消极的主题,也要试图以积极专业的方式结束这封信函。

4）正式的商务信函使用传统的结尾

商务信函中最常见的谦称结尾是真诚地(Sincerely)。其他的都可以基于此进行变化,如你真诚地(Sincerely yours)。使用恭敬地(Respectfully)这个词结尾可以表达你对读者的敬意,所以在敬意这种情况下使用这样的结尾。

5）非正式信件使用个性化结尾

对于写给朋友和熟人的个人或非正式的信函,可以使用如亲切地(Cordially)、热情的问候(Warm regards)和最美好的祝愿(Best wishes)这类结束语。

6）在签名档中写明职位

在正式的商务信函中,需要在名字旁写上头衔或职位。一个好的经验法则是如果你给收件人加了头衔或职位,那么也要列出你的职位。

7）在签名档中写明公司

如果你是公司的代理,比如提交一个方案或合同,在谦称结尾和四行之后的签名下面要写明公司的法定名称。这表明你是公司的代表,不是代表个人。

8）提供其他描述

在适当的时候,包括附件(Enclosure 或 Enc)表明你随函寄了一些材料。如果是你写的信,但是是别人打的字,还要包括作为证明人的姓名首字母。例如,KL:MCD 表明 KL 写信,MCD 打字。

5．日常信函撰写

虽然可以使用分块的方式来撰写正式的商务信函,但也可以使用一个简化的信函格式来撰写日常信函,使用这样更直接、不那么正式的方式可以方便地发送大量信函,如销售信函,以及将通知发送给客户、股东、供应商或员工。这样简化的信函格式省略了称呼、结尾谦称和签名,更专注于第一行和信的正文,如图 8-4 和图 8-5 所示。

1）用主题行替换称呼语

信的开始就要说明这封信的主题,强调目的,这样读者才可以立即预测和了解其余部分。

2）第一行说明目的

提出信中提出的提议、请求、应答、要解决的问题,或正在采取行动的明确的声明。

3）正文中详细描述

在正文部分可以阐述能够支持陈述在第一行的目的的一些细节。这些细节可以提供有关的提议或请求,列出想法的优势,或提供相关的事实。按照逻辑来安排这些信息,如按时间顺序或按照重要程度排序。直接称呼读者为你(you),并且关注于如何使信的内容能够让读者受益。

4）调整正文段落的格式增加可读性

使用尽可能少的段落。取而代之的是,使用编号或项目符号列表、表格、图形可以使信

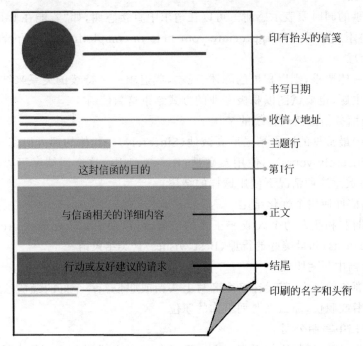

图 8-4 简化的信函格式的大纲

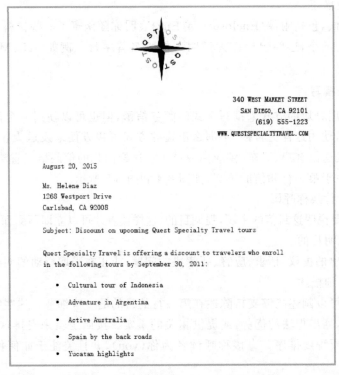

图 8-5 简化了格式的 Quest 公司的信函

函更容易阅读。

5）省略敬意的结尾

简化的信函格式并不适用于正式的信函,也就不需要一个结尾。取而代之的是,应该写一个结尾段来进行总结。

6. 放弃签名

在简化的信函格式中手写签名是不需要的。在许多情况下,可能会发送相当多份信函,那样在每一封上签名会很麻烦。如果使用彩色打印机来打印信件,可以选择蓝色的墨水来打印签名选项。

# Part 3

# Listening & Speaking

## Dialogue: Reinforcement Learning

在线音频

(*Because reinforcement learning is related with machine learning, Henry, Mark and Sophie are having a discussion about it.*)

Henry: As I know, reinforcement learning has some relationships with machine learning, right?

Mark: That is right. Reinforcement Learning (RL) is an area of machine learning concerned with how software agents ought to take actions in an environment so as to maximize some **notion** of cumulative reward.

Sophie: And the problem, due to [1] its generality, is studied in many other disciplines, such as game theory, control theory, operations research, information theory, simulation-based optimization, multi-agent systems, swarm intelligence, statistics and genetic algorithms. In the operations research and control **literature**, reinforcement learning is called approximate dynamic programming, or neuro-dynamic programming.

Mark: I agree with you. The problems of interest in reinforcement learning have also been studied in the theory of optimal control, which is concerned mostly with the existence and characterization of optimal solutions, and algorithms for their exact computation,

[1] Replace with:
1. because of
2. owning to
3. on account of

and less with learning or **approximation**, particularly in the absence of a mathematical model of the environment. In economics and game theory, reinforcement learning may be used to explain how **equilibrium** may arise under **bounded rationality**.

Henry: So, what is the difference between machine learning and reinforcement learning?

Mark: Well, in machine learning, the environment is typically formulated as a Markov Decision Process (MDP), as many reinforcement learning algorithms for this context utilize dynamic programming techniques. The main difference between the classical dynamic programming methods and reinforcement learning algorithms is that the latter do not assume knowledge of an exact mathematical model of the MDP and they target large MDPs where exact methods become infeasible.

Sophie: Moreover, reinforcement learning is considered as one of three machine learning paradigms, alongside supervised learning and unsupervised learning.

Henry: I guess there must be a difference between reinforcement learning and supervised learning.

Mark: Of course. Reinforcement learning differs from supervised learning in that [2] correct input/output pairs need not be presented, and sub-optimal actions need not be explicitly corrected. Instead the focus is on performance, which involves finding a balance between exploration (of **uncharted territory**) and exploitation (of current knowledge). The exploration vs. exploitation **trade-off** has been most thoroughly studied through the multi-armed bandit problem① and in finite MDPs.

Sophie: Thus, reinforcement learning is particularly **well-suited** to problems that include a long-term versus short-term reward trade-off. It has been applied successfully to various problems, including robot control, elevator scheduling, telecommunications, backgammon, checkers and go (AlphaGo).

[2] Replace with:
1. now that
2. because

# Unit 8  Reinforcement Learning

Henry: Which elements make reinforcement learning so powerful?

Sophie: Well, two elements make reinforcement learning powerful: the use of samples to optimize performance and the use of function approximation to deal with[3] large environments.

Mark: And thanks to these two key components, reinforcement learning can be used in large environments in the following situations: A model of the environment is known, but an analytic solution is not available; Only a simulation model of the environment is given (the subject of simulation-based optimization); the only way to collect information about the environment is to interact with it.

Sophie: Yes, the first two of these problems could be considered planning problems (since some form of model is available), while the last one could be considered to be a genuine learning problem. However,[4] reinforcement learning converts both planning problems to machine learning problems.

Mark: That's right!

[3] Replace with:
1. cope with
2. handle
3. address

[4] Replace with:
1. Nevertheless,
2. Yet,

## Exercises

Work in a group, and make up a similar conversation by replacing the statements with other expressions on the right side.

## Words

notion ['nəuʃn] n. 概念,想法,意图
literature ['lɪtrətʃə(r)] n. 文献
approximation [əˌprɒksɪ'meɪʃn] n. 近似法

equilibrium [ˌiːkwɪ'lɪbriəm; ˌekwɪ'lɪbriəm] n. 平衡,均衡
trade-off  权衡,取舍
well-suited  便利的,适当的

## Phrases

bounded rationality  有限理性
uncharted territory  未知的领域

153

## Notes

① 多臂老虎机问题（Multi-armed bandit problem）是这样的一种赌博工具：
- 它有多个可操作的拉杆；
- 操作每个拉杆所得的收益是随机的，但是每个拉杆收益的平均值（期望值）并不相同，有些高，有些低；
- 赌博人对于这个老虎机毫无了解，而且也无法通过观察别人的操作来了解（因为没有别人），只能自己去探索；
- 诚然，问题的核心是赌博人想在多次尝试的情况下获得最大化的收益。

## Listening Comprehension：Deep Reinforcement Learning

*Listen to the article and answer the following 3 questions based on it . After you hear a question , there will be a break of 15 seconds . During the break , you will decide which one is the best answer among the four choices marked （A）,（B）,（C） and （D）.*

**Questions**

1. What is the name of combination of RL with deep learning techniques? （　　）
   - （A） RL deep learning
   - （B） Deep RL
   - （C） Deep learning RL
   - （D） None of the above

2. How many games are mentioned where DeepMind was able to learn how to play? （　　）
   - （A） One
   - （B） Two
   - （C） Three
   - （D） Four

3. Which of the following is not right about DeepMind? （　　）
   - （A） DeepMind is a subsidiary of Alphabet
   - （B） DeepMind is a British company
   - （C） Alphabet is a subsidiary of DeepMind
   - （D） None of the above

## Words

| | |
|---|---|
| address [əˈdres] v. 处理（问题） | tough [tʌf] adj. 坚强的，不屈不挠的 |

154

Unit 8　Reinforcement Learning

## Phrases

with respect to　关于，至于
stand out　突出
in response　作为回答

## Dictation：Reinforcement Learning Challenges

在线音频

*This article will be played three times. Listen carefully, and fill in the numbered spaces with the appropriate words you have heard.*

As ___1___ as Reinforcement Learning（RL）has become in recent years, there are a number of ___2___ with using the technique. The main challenge is the preparation of the ___3___ environment which ___4___ to be highly dependent on the ___5___ task to be performed. When using the model for familiar game environments like Atari, Chess or Go, ___6___ the simulation environment is ___7___ simple. On the other hand, when it comes to building a model capable of driving an ___8___ vehicle, building an accurate simulator is ___9___ before turning the car ___10___ on public streets. The model has to determine out how to brake and avoid ___11___ in a safe environment. Moving out of the training environment and into to the real world is where things get ___12___.

___13___ and **fine-tuning** the neural network controlling the agent is another challenge. The only way to **converse** with the network is ___14___ the system of rewards and ___15___. This process may lead to something called ___16___ forgetting, where acquiring new knowledge causes some of the old to be ___17___ from the network. Yet another challenge is reaching a local ___18___, i.e. the agent performs the task as it is, but not necessarily in the ___19___ or required way. Lastly, there are agents that will ___20___ for realizing the goal without performing the task for which it was designed.

## Words

fine-tune[ˌfaɪn ˈtjuːn] v. 调整　　　　converse [kənˈvɜːs；ˈkɒn-] v. 交谈，谈话

# Unit 9

## Computer Vision

Unit 9  Computer Vision

> Part 1

# Reading & Translating

## Section A: What's the Difference between Computer Vision, Image Processing and Machine Learning?

Computer vision, image processing, signal processing, machine learning - you've heard the terms but what's the difference between them? Each of these fields is based on the input of an image or signal. They process the signal and then give us altered output in return. So what distinguishes these fields from each other? The boundaries between these domains may seem obvious since their names already imply their goals and methodologies. However, these fields **draw heavily from** the methodologies of one another, which can make the boundaries between them **blurry**. In this article we'll draw the distinction between the fields according to the type of input used, and more importantly, the methodologies and outputs that **characterize** each one.

Let's start by defining the input used in each field. Many, if not all, inputs can be thought of as a type of a signal. We favor the engineering definition of a signal, that is, a signal is a sequence of discrete measurable observations obtained using a capturing device, be it a camera, a radar, ultrasound, a microphone, **et cetera**… (Figure 9-1). The dimensionality of the input signal gives us the first distinction between the fields. Mono-channel sound waves can be thought of as a one-dimensional signal of **amplitude** over time, **whereas** pictures are a two-dimensional signal, made up of rows and columns of pixels. Recording consecutive images over time produces video which can be thought of as a three-dimensional signal.

Figure 9-1  Head movement **vestibulogram** signal captured by low noise camera

Input of one form can sometimes be transformed to another. For example, ultrasound images are recorded using the reflection of sound waves from the object observed, and then transformed to a visual **modality**. X-ray can be considered similarly to ultrasound, and only that radioactive **absorption** is transformed into an image. Magnetic

Resonance Imaging (MRI), records the **excitation**[1] of ions and transforms it into a visual image. In this sense, signal processing might be understood actually as image processing.

Let's look at the x-ray as a prototypical example. Let's assume we have acquired a single image from an x-ray machine. Image processing engineers (or software) would often have to improve the quality of the image before it passes to the physician's display. Hence, the input is an image and the output is an image. Image processing is, as its name implies, **all about** the processing of images. Both the input and the output are images. Methods frequently used in image processing are: filtering, noise removal, edge detection (Figure 9-2), color processing and **so forth**. Software packages **dedicated to** image processing are, for example, Photoshop and GIMP[2].

Figure 9-2　Edge detection in image processing software

In computer vision we wish to receive quantitative and qualitative information from visual data. **Much like** the process of visual reasoning of human vision, we can distinguish between objects, classify them, sort them according to their size, and so forth. Computer vision, like image processing, takes images as input. However, it returns another type of output, namely information on size, color, number, et cetera. Image processing methods are harnessed for achieving tasks of computer vision.

Extending beyond a single image, in computer vision we try to extract information from video. For example, we may want to count the number of cats passing by a certain point in the street as recorded by a video camera. Or, we may want to measure the distance run by a soccer player during the game and extract other statistics. Therefore, temporal information plays a major role in computer vision, much like it is with our own way of understanding the world.

But not all processes are understood **to their fullest**, which hinders our ability to construct a reliable and well-defined algorithm for our tasks. Machine learning methods then **come to our rescue**. Methodologies like Support Vector Machine (SVM) and Neural Networks are aimed at mimicking our way of reasoning without having full knowledge of how we do this. For example, a sonar machine placed to alert for intruders in oil drill facilities at sea needs to be able to detect a single diver in the **vicinity** of the facility. By sonar alone it is not possible to detect the difference between a big fish and a diver— more in-depth analysis is needed(Figure 9-3).

Figure 9-3  Fish or Diver? Image is by OpenCago.info—Licensed under CC BY-SA 2.5.

Characterizing the difference between the motions of the diver compared to a fish by sonar would be a good start. Features related to this motion, such as frequency, speed and so on are fed into the Support Vector Machine or neural network classifier. With training, the classifier learns to distinguish a diver from a fish. After the training set is completed, the classifier is intended to repeat the same observation as the human expert will make in a new situation. Thus, machine learning is quite a general framework in terms of input and output. Like humans, it can receive any signal as an input and give almost any type of output.

The following table (Table 9-1) summarizes the input and output of each domain:

Table 9-1  Input and output of each domain

| Domain | Input | Output |
| --- | --- | --- |
| Image processing | Image | Image |
| Signal processing | Signal | Signal, quantitative information, e.g. Peak location |
| Computer vision | Image/video | Image, quantitative/qualitative information, e.g. size, color, shape, classification, etc. |
| Machine learning | Any feature signal, from e.g. image, video, sound, etc. | Signal, quantitative/qualitative information, image… |

This can also be presented in a Venn diagram (Figure 9-4).

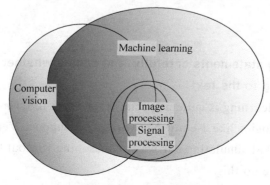

Figure 9-4  Venn diagram

## Words

heavily [ˈhevili] adv. 在很大程度上
blurry [ˈblɜːri] adj. 模糊的,不清楚的
characterize [ˈkærəktəraiz] v. 特征化,表示,表现
amplitude [ˈæmplitjuːd] n. 振幅,幅度
whereas [ˌweərˈæz] conj. 然而,鉴于
vestibulogram 针对前庭眼球震颤评估的图形记录
modality [məʊˈdæləti] n. 形式,形态
absorption [əbˈzɔːpʃn] n. (液体、气体等的) 吸收
resonance [ˈrezənəns] n. 共振
excitation [ˌeksaɪˈteɪʃ(ə)n] n. 励磁,激发
vicinity [vəˈsɪnəti] n. 邻近,附近

## Phrases

draw from  从……中得到,从……提取
et cetera  等等
all about  到处,各处
so forth  等等
dedicated to  致力于
much like  就像,很像
to one's fullest  充分地,达到最大程度
come to one's rescue  救援,前来营救

## Notes

[1] 励磁(excitation)就是向发电机或者同步电动机定子提供定子电源,为发电机等(利用电磁感应原理工作的电气设备)提供工作磁场的机器。有时向发电机转子提供转子电源的装置也叫励磁。

[2] Gimp(GNU image manipulation program, GNU 图像处理程序),它是一个图像处理与合成工具。Gimp 的扩展性很强,用户可以通过自己编写的插件来扩充 Gimp 功能。

## Exercises

I. Read the following statements carefully, and decide whether they are true (T) or false (F) according to the text.

  ____ 1. Machine learning is quite a special framework in terms of input and output.
  ____ 2. Methodologies like Neural Networks and Support Vector Machine (SVM) are aimed at simulating our way of reasoning without having full knowledge of how we do this.
  ____ 3. Recording consecutive images over time produces video which can be

thought of as a one-dimensional signal.

____ 4. Computer vision, image processing, signal processing, machine learning, each of these fields is based on the input of an image or signal.

____ 5. A signal is a sequence of discrete measurable observations obtained using a capturing device in terms of engineering definition of a signal.

II. **Choose the best answer to each of the following questions according to the text.**

1. Which of the following is not mentioned in the Venn diagram? (    )
   A. Machine learning
   B. Computer vision
   C. Input and output
   D. Image processing

2. Which of the following is not right about machine learning? (    )
   A. Machine learning is quite a general framework in terms of input and output.
   B. Machine learning can receive any signal as an input and give almost any type of output.
   C. Machine learning methods like Support Vector Machine (SVM) and Neural Networks are aimed at mimicking our way of reasoning without having full knowledge of how we do this.
   D. Machine learning hinders our ability to construct a reliable and well-defined algorithm for our tasks.

3. Which of the following is right about computer vision? (    )
   A. Computer vision takes images as input.
   B. Computer vision returns another type of output, namely information on size, color, number, et cetera.
   C. Image processing methods are harnessed for achieving tasks of computer vision.
   D. All of the above

III. **Fill in the numbered spaces with the words or phrases chosen from the box. Change the forms where necessary.**

> effective  use  process  manage  give  facilitate
> by  need  characteristic  describe

**Feature Extraction**

Feature extraction is a process of dimensionality reduction ___1___ which an initial set of raw data is reduced to more ___2___ groups for processing. A ___3___ of these large data sets is a large number of variables that require a lot of computing resources to process. Feature extraction is the name for methods that select and/or combine variables into features, ___4___ reducing the amount of data that must be processed, while still accurately and completely ___5___ the original data set.

The process of feature extraction is ___6___ when you need to reduce the number of resources ___7___ for processing without losing important or relevant information. Feature extraction can also reduce the amount of redundant data for a ___8___ analysis. Also, the reduction of the data and the machine's efforts in building variable combinations (features) ___9___ the speed of learning and generalization steps in the machine learning ___10___.

IV. Translate the following passage into Chinese.

**Pattern Recognition**

Pattern recognition is the process of recognizing patterns by using machine learning algorithm. Pattern recognition can be defined as the classification of data based on knowledge already gained or on statistical information extracted from patterns and/or their representation. One of the important aspects of the pattern recognition is its application potential.

In a typical pattern recognition application, the raw data is processed and converted into a form that is amenable for a machine to use. Pattern recognition involves classification and cluster of patterns.

## Section B: Using Vision for Controlling Movement

One of the principal uses of vision is to provide information both for manipulating objects—picking them up, grasping them, **twirling** them, and so on—and for navigating while avoiding obstacles. The ability to use vision for these purposes is present in the most primitive of animal visual systems. In many cases, the visual system is minimal, in the sense that it extracts from the available light field just the information the animal needs to inform its behavior. Quite probably, modern vision systems evolved from early, primitive organisms that used a photosensitive spot at one end to orient themselves toward (or away from) the light.

Let us consider a vision system for an automated vehicle driving on a **freeway**. The tasks faced by the driver include the following:

1. **Lateral** control—ensure that the vehicle remains securely within its lane or changes lanes **smoothly** when required.

2. **Longitudinal** control—ensure that there is a safe distance to the vehicle in front.

3. Obstacle avoidance—monitor vehicles in neighboring lanes and be prepared for evasive maneuvers if one of them decides to change lanes.

The problem for the driver is to generate appropriate steering, acceleration, and braking actions to best accomplish these tasks.

For lateral control, one needs to maintain a representation of the position and orientation of the car relative to the lane. We can use edge-detection algorithms to find edges corresponding to the lane-marker segments. We can then fit smooth curves to these edge elements. The parameters of these curves carry information about the lateral position of the car, the direction it is pointing relative to the lane, and the **curvature** of the lane. This information, along with information about the dynamics of the car, is all that is needed by the steering-control system. If we have good detailed maps of the road, then the vision system **serves** to confirm our position (and to watch for obstacles that are not on the map).

For longitudinal control, one needs to know distances to the vehicles in front. This can be accomplished with **binocular stereopsis** or optical flow. Using these techniques, vision controlled cars can now drive reliably at highway speeds.

The more general case of mobile robots navigating in various indoor and outdoor environments has been studied, too. One particular problem, localizing the robot in its environment, now has pretty good solutions. A group at Sarnoff [1] has developed a system based on two cameras looking forward that track feature points in 3D and use that to reconstruct the position of the robot relative to the environment. In fact, they have two **stereoscopic** camera systems, one looking front and one looking back—this gives greater robustness in case the robot has to go through a **featureless patch** due to dark shadows, **blank walls**, and **the like** (Figure 9-5). It is unlikely that there are no features either in the front or in the back. Now of course, that could happen, so a **backup** is provided by using an **Inertial** Motion Unit (IMU) somewhat **akin to** the mechanisms for sensing acceleration that we humans have in our inner ears. By integrating the sensed acceleration twice, one can keep track of the change in position. Combining the data from vision and the IMU is a problem of probabilistic evidence **fusion** and can be tackled using techniques, such as Kalman filtering [2].

Figure 9-5  Stereoscopic camera systems

In the use of visual **odometry** (estimation of change in position), as in other problems of odometry, there is the problem of "drift," positional errors accumulating over time. The solution for this is to use landmarks to provide absolute position fixes: as soon as the robot passes a location in its internal map, it can adjust its estimate of its position appropriately. Accuracies on the order of centimeters have been demonstrated with these techniques.

The driving example makes one point very clear: for a specific task, one does not need to recover all the information that, in principle, can be recovered from an image. One does not need to recover the exact shape of every vehicle, solve for shape-from-texture on the grass surface **adjacent to** the freeway, and so on. Instead, a vision system should compute just what is needed to accomplish the task.

 **Words**

twirl [twɜːl] v. (使)旋转，转动
freeway [ˈfriːweɪ] n. 高速公路
lateral [ˈlætərəl] adj. 横向的，侧面的
smoothly [ˈsmuːðli] adv. 平稳地，顺利地
longitudinal [ˌlɒŋɡɪˈtjuːdɪnl；ˌlɒndʒɪˈtjuːdɪnl] adj. 纵向的
curvature [ˈkɜːvətʃə(r)] n. 弯曲，[数]曲率
serve [sɜːv] v. 对……有用，可作……用

stereoscopic [ˌsteriəˈskɒpɪk] adj. 立体的，实体镜的
featureless [ˈfiːtʃələs] adj. 一般的，平淡无奇的
patch [pætʃ] n. 小块土地
backup [ˈbækʌp] n. 备份，支持
inertial [ɪˈnɜːʃl] adj. 惯性的，不活泼的
fusion [ˈfjuːʒn] n. 融合，熔化
odometry [ɒˈdɒmɪtri] n. 量距，测程法

 **Phrases**

binocular stereopsis　双目立体视觉
blank wall　无法克服的障碍
the like　类似的东西
akin to　类似于，同类
adjacent to　临近的，靠近的

 **Notes**

[1] Sarnoff 公司是美国国际集团（AIG）的子公司，主要开发定制微型数码相机。这种定制的相机必须提供迅速曝光调整的功能，且对光线突然改变有所反应。此外，这种新数码相机技术还具备最先进的动态范围功能，不论是在明亮的日光或在黑暗的阴影中，都能捕捉影像的细微部分；它的主动像素感应技术提供的动态范围功能，比普通相机高 100 倍；这种技术的耗电量约 600mW，约是其他技术的 1/5。

[2] 卡尔曼滤波（Kalman filtering）是一种利用线性系统状态方程，通过系统输入输出观测数据，对系统状态进行最优估计的算法。由于观测数据中包括系统中的噪声和干扰的影响，所以最优估计也可看作是滤波过程。

 **Exercises**

I. Read the following statements carefully, and decide whether they are true (T) or false (F) according to the text.

　　____ 1. Sarnoff is a plane company's name.
　　____ 2. Lateral control ensures that there is a safe distance to the vehicle in front.
　　____ 3. Longitudinal control ensures that the vehicle remains securely within its lane

or changes lanes smoothly when required.

____ 4. Longitudinal control monitors vehicles in neighboring lanes and be prepared for evasive maneuvers if one of them decides to change lanes.

____ 5. Kalman filtering is a technique.

II. **Choose the best answer to each of the following questions according to the text.**

1. Which of the following company is mentioned? (　　)
   A. Microsoft
   B. Sarnoff
   C. IBM
   D. Oracle

2. How many tasks faced by the driver who drives an automated vehicle using a vision system on a freeway? (　　)
   A. One
   B. Two
   C. Three
   D. Four

3. Which of the following algorithm can be used to find edges corresponding to the lane-marker segments? (　　)
   A. Kalman filtering
   B. KNN
   C. SVM
   D. Edge-detection

III. **Fill in the blanks with the words or phrases chosen from the box. Change the forms where necessary.**

> network　like　approach　train　give　use
> relate　together　identify　distinguish

**Computer Vision Resembles a Jigsaw Puzzle**

Computers assemble visual images in the same way you might put ___1___ a jigsaw puzzle.

Think about how you ___2___ a jigsaw puzzle. You have all these pieces, and you need to assemble them into an image. That's how neural ___3___ for computer vision work. They ___4___ many different pieces of the image, they ___5___ the edges and then model the subcomponents. ___6___ filtering and a series of actions through deep network layers, they can piece all the parts of the image together, much ___7___ you would with a puzzle.

The computer isn't ___8___ a final image on the top of a puzzle box — but is often

165

fed hundreds or thousands of _____9_____ images to train it to recognize specific objects.

Instead of _____10_____ computers to look for whiskers, tails and pointy ears to recognize a cat, programmers upload millions of photos of cats, and then the model learns on its own the different features that make up a cat.

IV. Translate the following passage into Chinese.

**Image Segmentation in Deep Learning**

Many computer vision tasks require intelligent segmentation of an image, to understand what is in the image and enable easier analysis of each part. Today's image segmentation techniques use models of deep learning for computer vision to understand, at a level unimaginable only a decade ago, exactly which real-world object is represented by each pixel of an image.

Deep learning can learn patterns in visual inputs in order to predict object classes that make up an image. The main deep learning architecture used for image processing is a Convolutional Neural Network (CNN), or specific CNN frameworks like AlexNet, VGG, Inception, and ResNet. Models of deep learning for computer vision are typically trained and executed on specialized Graphics Processing Units (GPUs) to reduce computation time.

# Part 2

# Simulated Writing: Writing for Employment (I)

有效地搜寻就业机会是可以培养的最重要的职业技能之一。求职的时候需要使用所有的沟通技巧，其中大部分的早期步骤涉及书面交流。那些按时提交了专业文档并且在简历上看起来很不错的申请人往往是雇主邀请进入最后面试的人。

1. 了解求职

找到合适工作的第一步是评估你的兴趣、目标和资历，然后确定相应的工作和雇主。求职将一直持续到最后一步——接到录用通知。研究表明，一般人会在一生中改变从业领域三到五次，并在每个从业领域都有数次工作机会，因此发展求职能力将会为你提供十分有价值的生存技能。求职的细节将会依照每份工作有所变化，但是如图9-6所示的一般步骤是保持不变的。

大多数求职过程都包括以下步骤：

1) 确定就业目标

在完成第一份申请之前，对要寻找什么样的工作要有一个清晰的概念。可以问自己一些问题，比如说：你的专业技能是什么？天赋和兴趣点在哪里？你现在想要寻找什么样的职位？这个职位和你的长期职业目标契合吗？在你感兴趣的领域中求职市场的特性是什么？你想要生活在什么样的环境之下呢？参见图9-7，你的回答将帮助你制定专属于你的求职策略。

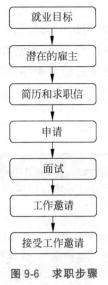

图9-6 求职步骤

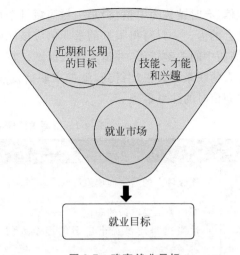

图9-7 确定就业目标

2）识别潜在的雇主

待选的工作数目取决于联系的雇主的数量。想要识别潜在的雇主，就要善于利用网站和平面媒体提供的就业信息和岗位空缺，并且要花费大量的时间与他人（朋友、熟人、老师和家人等）交谈就业机会。在个人通信的网络（包括社交网络）上进行交流，是找到工作机会以及发现潜在雇主的有价值信息的最有效方法。

3）编写有针对性的简历和求职信

在申请特定职位时，简历和求职信将成为可以用来提升自己的营销工具。有效的简历和求职信就像是为想要申请的岗位特别定制的一般，并且它会突出相关的资历。有一些求职者会准备几份针对性不同的简历，并利用它们在不同的行业和岗位来进行申请。

4）申请合适的职位

雇主需要一个可以从中找到能够填补空缺职位的申请者的候选群体。很大程度上他们会从符合条件和有资格的人当中来挑选，但还是由你来标明你的兴趣。有一些组织会要求填写一份申请表，还有一些组织会选择接受简历和求职信。应该确定每一份申请的流程，并且认真仔细地开展每一个步骤。

5）为面试做准备

递交简历和求职信的目的是得到面试的机会。面试是和招聘经理既要讨论工作机会又要讨论你的资历的一次会面。许多机构的初步面试是通过电话进行的，但基本不会在没有进行一个面对面的会面之前做出聘用的决定。最后得到这份工作的人往往是为面试做了最好的准备的人——并不总是一个最能胜任这个位置的人。

6）接受工作邀请

接受一个满足就业目标的就业邀请是实现终极目标的最后一步。得到工作邀请后，要对这个工作做个评估，看它是否符合你的需求和兴趣。最好的工作邀请是同时非常适合你和你的雇主的。仔细权衡得到的工作邀请的优劣，接受那些符合你的兴趣、天赋和求职目标的邀请。

**2. 撰写有效的求职信**

求职信(有时被称为申请信)是一封简短且个性化的、伴随着简历一起发送给你感兴趣的雇佣机会的信。一封写得很好的求职信应该能够吸引读者的注意力,标明想要申请的职位,突出简历中的关键资历,并请求进入面试。将求职信和简历发送给负责招聘决策的人,如果这个人的联系方式很难得到,就把求职信发送到这个组织的人力资源部门。表 9-1 列出了写求职信时的注意事项。

表 9-1 撰写求职信的注意事项

| 求职信元素 | 适合提到 | 尽量避免 |
| --- | --- | --- |
| 问候 | • 用收件人的名字向他致辞 | • 使用笼统的称呼来进行问候,比如尊敬的人事部经理(Dear Hiring Manager) |
| 开头 | • 明确表示要申请的职位<br>• 适当提到建议提出申请的那个人的名字<br>• 说明是从哪里看见工作招聘广告的 | • 笼统地指向众多职位,例如"我想要申请一个贵公司现在空缺的职位"(I'm applying to fill the job vacancy at your company.) |
| 资历 | • 列举三至五个与职位直接相关的资历<br>• 描述可以为雇佣者做些什么<br>• 重点表现优势可以如何使雇主受益 | • 列举所学课程或者之前的工作职责<br>• 包括不具备的技能和经历<br>• 描述资历中的各种细节或者重复简历上的细节 |
| 总结 | • 自信地请求一次面试<br>• 如果没有提到简历,那么就在这部分提起<br>• 提供电话号码并说明最方便的联系时间 | • 直接地要求获得这份工作<br>• 让求职信看起来不真诚、要求过高或者过分担忧 |
| 后续跟进 | 如果没有收到一个潜在雇佣者的回信,那么就将求职信和简历重发一次 | 忽略来自雇佣者的关于后续跟进的建议和指导 |

1) 用收件人的名字向他们致辞

在开始写求职信的时候,用收件人的名字向他们致辞。这会吸引他们的注意力,并且使申请显得更加人性化。避免使用通用的称呼,例如"尊敬的先生/女士"(Dear Sir/Madam)或"敬启者"(To Whom It May Concern)。

2) 确定申请职位

当进行申请的时候,目标单位可能有几个职位空缺,在求职信的第一段应该确定想要寻求的那个职位。如果时机适宜,可以提一下是谁建议提出申请或者是从哪里看到这个工作广告的。

3) 突出最相关的技能

在求职信的第二段应简要介绍三到五个就职资格、所获成就或与这个空缺职位相关的技能。推销自己的长处来显示能为雇主做些什么,但不必细讲。目标是引起对方的兴趣,并鼓动他们认真阅读简历。当描述到资历的时候,可以建议其参考简历来获取更多的信息。

图 9-8 是一封求职信的开始部分。

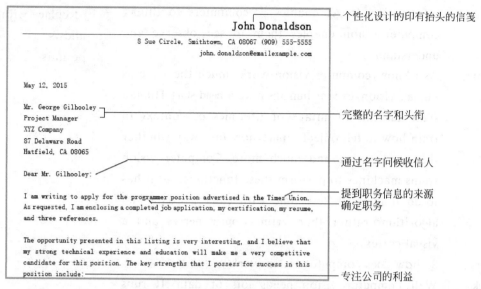

图 9-8　修订过的求职信的开始部分

4）用请求行动来进行总结

求职信和简历的直接目标是让申请通过初步的筛选，并且能够得到面试机会，那么就可以通过直接请求面试机会来结束求职信。

5）用令人关注的期望来跟进

最初的申请并不总能在正确的时间被正确的人看到，如果在几周之内都没有收到任何回应就再发送一次信件。对申请状况进行持续跟进会向招聘经理显示主动性以及对这份工作机会的热情。

转 184 页

## Part 3

# Listening & Speaking

### Dialogue：Computer Vision

在线音频

（Computer vision is one of the hottest area of computer science and artificial intelligence research. Sophie，Henry and Mark are talking about it.）

Henry：　Do you know what computer vision is?

Mark：　Yes. Computer vision is a field of Artificial Intelligence（AI）that enables computers and systems to derive meaningful information from digital images，videos and other visual inputs — and take actions or make recommendations based on that information. If AI enables [1] computers to think, computer vision enables them to see, observe and understand.

[1] Replace with:
1. allows
2. permits

Sophie: As I know, computer vision works much the same as human vision, except humans have a **head start**. Human sight has the advantage of lifetimes of context to train how to **tell** objects **apart**, how far away whether they are moving and much more. Computer vision trains machines to perform these functions, but it has to do it in much less time with cameras, data and algorithms rather than **retinas**, **optic nerves** and a visual cortex.

Henry: So how does computer vision work?

Mark: Well, computer vision needs lots of data. It runs analyses of data **over and over** until it discerns distinctions and ultimately recognizes images. For example, to train a computer to recognize apples, it needs to be fed vast quantities of apple images and apple-related items to learn the differences and recognize an apple.

Sophie: Here two essential technologies are used to accomplish this: a type of machine learning called deep learning and a Convolutional Neural Network (CNN).

Mark: Right. Machine learning uses algorithmic models that enable a computer to teach itself about the context of visual data. If enough data is fed through the model, the computer will "look" at the data and teach itself to tell one image from another. Algorithms enable the machine to learn by itself, rather than someone programming it to recognize an image.

Sophie: And a CNN helps a machine learning or deep learning model "look" by breaking images down into pixels that are given tags or labels. It uses the labels to perform convolutions (a mathematical operation on two functions to produce a third function) and makes predictions about what it is "seeing." The neural network runs convolutions and checks the accuracy

|  | of its predictions in a series of [2] iterations until the predictions start to come true. It is then recognizing or seeing images in a way similar to humans. |
|---|---|
| Henry: | So interesting. |
| Mark: | And also much like a human **making out** an image **at a distance**, a CNN first discerns hard edges and simple shapes, and then fills in information as it runs iterations of its predictions. A CNN is used to understand single images. A Recurrent Neural Network (RNN) is used in a similar way for video applications to help computers understand how pictures in a series of **frames** are related to one another. |
| Henry: | Ok. Could you please tell me why computer vision is important? |
| Sophie: | Well, there is a lot of research being done in the computer vision field, but it's not just research. Real-world applications demonstrate how important computer vision is to **endeavor in** business, entertainment, transportation, healthcare and everyday life. |
| Mark: | Right. A key driver for the growth of these applications is the **flood** of visual information flowing from smartphones, security systems, traffic cameras and other visually **instrumented** devices. The information creates a **test bed** to train computer vision applications. |
| Henry: | Quite exciting. Very nice to talk with you. See you. |
| Sophie & Mark: | Me too. See you. |

[2] Replace with:
1. a range of
2. an array of
3. a set of

## Exercises

Work in a group, and make up a similar conversation by replacing the statements with other expressions on the right side.

## Words

retina['retinə] n. 视网膜
frame[freim] n. 帧, 画面
flood[flʌd] n. 一大批

instrument['instrəmənt] v. 给……装测量仪器

## Phrases

head start　抢先起步的优势,有利的开端
tell apart　区分,分辨
optic nerve　视神经
over and over　反复,再三
make out　理解,辨认出
at a distance　在远处,有相当距离
endeavor in　在……的努力
test bed　试验台

## Listening Comprehension：Pattern Recognition

*Listen to the article and answer the following 3 questions based on it. After you hear a question, there will be a break of 15 seconds. During the break, you will decide which one is the best answer among the four choices marked（A）,（B）,（C）and（D）.*

**Questions**

1. What does KDD stand for according to this article?（　　）
   (A) Knowledge Discriminant Data
   (B) Knowledge Dedicated Data
   (C) Knowledge Discovery in Databases
   (D) Knowledge Discount Data

2. Which of the following is right?（　　）
   (A) Pattern recognition is closely related to artificial intelligence.
   (B) Pattern recognition is closely related to machine learning.
   (C) Pattern recognition is closely related to applications such as data mining and Knowledge Discovery in Databases（KDD）.
   (D) All of the above

3. In the last paragraph, which of the following algorithms are compared?（　　）
   (A) Pattern recognition algorithms and SVM
   (B) Pattern recognition algorithms and decision tree
   (C) Pattern recognition algorithms and pattern-matching algorithms
   (D) Pattern-matching algorithms and LDA

## Words

hand-crafted　手工制作的

Unit 9　Computer Vision

take into account　考虑,顾及
be opposed to　与……相对,反对……
textual data　文本数据

### Dictation: Artificial Intelligence, Machine Learning, Deep Learning, and Computer Vision… What is the Difference?

在线音频

　　This article will be played three times. Listen carefully, and fill in the numbered spaces with the appropriate words you have heard.

　　Artificial Intelligence is a ___1___ of Computer Science where machines can appear to be intelligent by running programs. It is a very ___2___ topic and covers everything from the automatic doors at the shopping mall to the most intelligent systems built today.

　　Machine Learning is the practice of giving a computer a set of ___3___ and tasks, then letting it **figure out** a way to complete those tasks. The machine **in essence starts out with** no knowledge and through ___4___ and error comes up with a ___5___ solution. The **work horse** of Machine Learning is the Neural Network.

　　Neural Networks are ___6___ and data structures designed to let machines classify and ___7___ outputs based on a series of inputs. The neural network is an ___8___ structure to the brain. It consists of Nodes (brain cells), connections, and ___9___ and works on the principle of **Gradient Descent**. The network has two ___10___ of operations: Training and Inference. In Training mode, lots of data sets are fed into the input nodes and the weights are ___11___. In Inference mode, the ___12___ data is fed into the input nodes and the system suggests an output. There is a lot more to understanding Neural Networks but this a very broad ___13___. Neural Networks are usually very complicated and take a lot of computing power to train.

　　Deep Learning Networks use Neural Networks **inside of** them. Deep Learning Networks and Neural Networks Architectures have a lot of things in ___14___. They both have an input and output layer and Training and Inference modes. But there are some new **twists** usually ___15___ in Deep Learning Networks like Convolution and Max Pooling [1] to make the algorithms run faster and allow for computation at great depths. **In a nutshell**, one can think of a Deep Learning Network as a network of Neural Networks.

　　Computer Vision is the ___16___ of giving machines knowledge of their physical surrounding world through sensors. In the past this was a very ___17___ and complicated task requiring a specific ___18___ algorithm to analyze ___19___. These algorithms were not flexible and had to be used in a specific case and were very **susceptible to** rotation and _____20_____. Recent developments in the speed and number of hardware GPUs have allowed Computer Vision to take advantage of Deep Learning Networks that help to **mitigate** the issues with experienced with standard computer vision algorithms.

 Words

| twist[twist] n.（故事或情况的）转折，转变 | mitigate['mitigeit] v.减轻,缓和,缓解 |

 Phrases

figure out  想出,理解,解决
in essence  本质上,其实
start out with  从……入手,开始
work horse  主力,主要设备
gradient descent  梯度下降
inside of  在……之内,少于
in a nutshell  简言之,概括地说
susceptible to  易受……影响的,对……敏感的

 Notes

［1］最大池化（max pooling）是在卷积后还会有一个池化的操作。它的操作是这样进行的：整个图片被不重叠地分割成若干个同样大小的小块（pooling size）；每个小块内只取最大的数字，再舍弃其他节点，保持原有的平面结构得到输出。最大池化和卷积核的操作区别：池化作用于图像中不重合的区域（这与卷积操作不同）。

# Unit 10

# Natural Language Processing

# Part 1

# Reading & Translating

## Section A: Language Understanding

One of the inherent capabilities of a human being is to understand—that is, interpret—the audio signals that they perceive. A machine that can understand natural language can be very useful in daily life. For example, it can replace—most of the time—a telephone operator. It can also be used on occasions when a system needs a predefined format of queries. For example, the queries given to a database must normally follow the format used by that specific system. A machine that can understand queries in natural language and translate them to formal queries can be very useful.

We can divide the task of a machine that understands natural language into four consecutive steps: speech recognition, syntactic analysis, semantic analysis, and **pragmatic analysis**.

The first step in natural language processing is speech recognition. In this step, a speech signal is analyzed and the sequence of words it contains is extracted. The input to the speech recognition subsystem is a continuous (analog) signal; the output is a sequence of words. The signal needs to be divided into different sounds, sometimes called **phonemes**. The sounds then need to be combined into words.

The syntactic analysis step is used to define how words are to be grouped in a sentence. This is a difficult task in a language like English, in which the function of a word in a sentence is not determined by its position in the sentence. For example, in the following two sentences:

Mary rewarded John.

John was rewarded by Mary.

It is always John who is rewarded, but in the first sentence John is in the last position and Mary is in the first position. A machine that hears any of the above sentences needs to interpret them correctly and come to the same conclusion no matter which sentence is heard.

The first tool to correctly analyze a sentence is a well-defined grammar. A fully developed language like English has a very long set of grammatical rules. We assume a very small subset of the English language and define a very small set of rules just to show the idea. The grammar of a language can be defined using several methods: we use a simple version of BNF (Backus-Naur Form) [1] that is used in computer science to define the syntax of a programming language (Table 10-1).

Table 10-1  A simple grammar

| | Rule | | |
|---|---|---|---|
| 1 | Sentence | → | NounPhrase VerbPhrase |
| 2 | Noun-Phrase | → | Noun\|Article Noun\|Article Adjective Noun |
| 3 | Verb Phrase | → | Verb\| Verb NounPhrase\| Verb NounPhrase Adverb |
| 4 | Noun | → | [home]\|[cat]\|[water]\|[dog]\|[John]\|[Mary] |
| 5 | Article | → | [a]\|[the] |
| 6 | Adjective | → | [big]\|[small]\|[tall]\|[short]\|[white]\|[black] |
| 7 | Verb | → | [goes]\|[comes]\|[eats]\|[drinks]\|[has]/[loves] |

The first rule defines a sentence as a noun **phrase** followed by a verb phrase. The second rule defines three choices for a noun phrase: a single noun, an **article** followed by a noun, or an article followed by an adjective and a noun. The fourth rule explicitly defines what a noun can be. In our simple language, we have defined only seven nouns; in a language like English the list of nouns is defined in a dictionary. The sixth rule also defines a very small set of adjectives and the seventh rule a small set of verbs.

Although the syntax of our language is very primitive, we can make many sentences out of it. For example, we can have:

John comes home.
Mary drinks water.
John has a white dog.
John loves Mary.
Mary loves John.

It should be clear that even a simple grammar as defined in Table 10-1 uses different options. A machine that determines if a sentence is grammatically (syntactically) correct does not need to check all possible choices before rejecting a sentence as an invalid one.

This is done by a **parser**. A parser creates a parse tree based on the grammar rules to determine the validity of a sentence. Figure 10-1 shows the parse tree for the sentence "John has a white dog." based on our rules defined in Table 10-1.

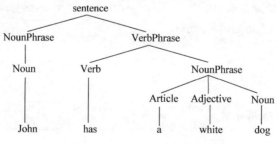

Figure 10-1  Parsing a sentence

The semantic analysis extracts the meaning of a sentence after it has been syntactically analyzed. This analysis creates a representation of the objects involved in the sentence, their relations, and their **attributes**. The analysis can use any of the knowledge representation schemes. For example, the sentence "John has a dog." can be represented using **predicate** logic as:

∃ x dog(x) has (John,x)

The three previous steps—speech recognition, syntax analysis, and semantic analysis—can create a knowledge representation of a spoken sentence. In most cases, another step, pragmatic analysis, is needed to further clarify the purpose of the sentence and to remove ambiguities.

The purpose of the sentence cannot be found using the three steps listed above. For example, the sentence "Can you swim a mile?" asks about the ability of hearer. However, the sentence 'Can you pass the salt?' is merely a polite request. An English language sentence can have many different purposes, such as informing, requesting, promising, inquiring, and so on. Pragmatic analysis is required to find the purpose of the sentence.

Sometimes a sentence is ambiguous after semantic analysis. Ambiguity can **manifest** itself in different ways. A word can have more than one function—for example, the word 'hard' can be used both as an adjective and an adverb. A word can also have more than one meaning— for example, the word 'ball' can mean different things in 'football' and 'ball room'. Two words with the same pronunciation can have different spellings and meanings— a sentence may be syntactically correct, but be **nonsense**. For example, the sentence 'John ate the mountain.' can be syntactically parsed as a valid sentence and be correctly analyzed by the semantic analyzer, but it is still nonsense. Another purpose of the pragmatic analyzer is to remove ambiguities from the knowledge representation sentence if possible.

## Words

| | |
|---|---|
| phoneme[ˈfəuni:m] n. 音素,音位 | predicate [ˈpredikət , ˈpredikeit] n. 谓词 |
| phrase[freiz] n. 词组,短语 | |
| article[ˈɑ:tikl] n. 冠词 | manifest[ˈmænifest] v. 表明 |
| parser[ˈpɑ:sə] n. 词法分析器 | nonsense[ˈnɔnsns] adj. 荒谬的 |
| attribute[əˈtribju:t,ˈætribju:t] n. 属性 | |

## Phrases

pragmatic analysis    语用分析

## Notes

[1] 巴科斯范式(Backus-Naur Form, BNF)以美国人巴科斯(Backus)和丹麦人诺尔(Naur)的名字命名的一种形式化的语法表示方法,用来描述语法的一种形式体系,是一种

典型的元语言。它不仅能严格地表示语法规则,而且所描述的语法是与上下文无关的。它具有语法简单,表示明确,便于语法分析和编译的特点。

 **Exercises**

I. Read the following statements carefully, and decide whether they are true (T) or false (F) according to the text.

    ____ 1. A sentence may be syntactically correct, but be nonsense.

    ____ 2. A parser creates a grammar tree based on the grammar rules to determine the validity of a sentence.

    ____ 3. The speech recognition, syntax analysis, and semantic analysis can create a knowledge representation of a spoken sentence.

    ____ 4. The syntax analysis extracts the meaning of a sentence after it has been semantically analyzed.

    ____ 5. The semantic analysis creates a representation of the objects involved in the sentence, their relations, and their attributes.

II. Choose the best answer to each of the following questions according to the text.

  1. How many consecutive steps can we divide the task of a machine that understands natural language into? (　　)

    A. One
    B. Two
    C. Three
    D. Four

  2. Which of the following step is used to define how words are to be grouped in a sentence? (　　)

    A. Speech recognition
    B. Syntax analysis
    C. Semantic analysis
    D. Pragmatic analysis

  3. Which of the following step extracts the meaning of a sentence after it has been syntactically analyzed? (　　)

    A. Speech recognition
    B. Syntax analysis
    C. Semantic analysis
    D. Pragmatic analysis

III. Fill in the blanks with the words or phrases chosen from the box. Change the forms where necessary.

> span　use　particular　create　retrieve　recognize
> random　require　general　refer

**Corpus**

One of the first things ___1___ for Natural Language Processing (NLP) tasks is a corpus. In linguistics and NLP, corpus (literally Latin for body) ___2___ to a collection of texts. Such collections may be formed of a single language of texts, or can ___3___ multiple languages—there are numerous reasons for which multilingual corpora (the plural of corpus) may be useful. Corpora may also consist of themed texts (historical, Biblical, etc.). Corpora are ___4___ solely used for statistical linguistic analysis and hypothesis testing.

A corpus provides grammarians, lexicographers, and other interested parties with better descriptions of a language. Computer-processable corpora allow linguists to adopt the principle of total accountability, ___5___ all the occurrences of a ___6___ word or structure for inspection or ___7___ selected samples. Corpus analysis provides lexical information, morphosyntactic information, semantic information and pragmatic information.

Corpora are ___8___ in the development of NLP tools. Applications include spell-checking, grammar-checking, speech ___9___, text-to-speech and speech-to-text synthesis, automatic abstraction and indexing, information retrieval and machine translation. Corpora are also used for ___10___ of new dictionaries and grammars for learners.

**IV. Translate the following passage into Chinese.**

**Machine Translation**

Machine translation systems are applications or online services that use machine-learning technologies to translate large amounts of text from and to any of their supported languages. The service translates a "source" text from one language to a different "target" language.

Although the concepts behind machine translation technology and the interfaces to use it are relatively simple, the science and technologies behind it are extremely complex and bring together several leading-edge technologies, in particular, deep learning (artificial intelligence), big data, linguistics, cloud computing, and Web APIs.

## Section B: Natural Language Processing vs. Machine Learning vs. Deep Learning

NLP, Machine Learning and Deep Learning are all parts of Artificial Intelligence (AI), which is a part of the greater field of Computer Science (CS). The following image (Figure 10-2) visually illustrates CS, AI and some of the components of AI:

- Robotics (AI for motion)
- Vision (AI for visual space—videos, images)
- NLP (AI for text)

There are other aspects of AI too which are

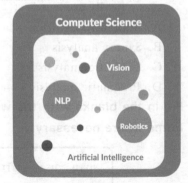

Figure 10-2  CS, AI and some of the components of AI

Unit 10　Natural Language Processing

not **highlighted** in the image—such as speech.

Natural Language Processing (or NLP) is an area that is a **confluence** of Artificial Intelligence and linguistics. It involves intelligent analysis of written language.

If you have a lot of data written in plain text and you want to automatically get some **insights** from it, you need to use NLP.

Some applications of NLP are:

• **Sentiment** analysis: Classification of emotion behind text content, e.g. movie reviews are good or bad. How can humans tell if a review is good or bad? Can we use the same features that humans use—presence of describing words (adjectives) such as "great" or "terrible" etc.?

• Information extraction: Extracting structured data from text, e.g. relationships between country and name of president, acquisition relationship between buyer and seller etc.

• Information retrieval: This is a synonym of search. It is the concept of retrieving the correct document given a query - like Google! For the curious, here is info on how to build your own search engine and some more details on the **internals** of Lucene (Apache Lucene is an open source search engine that is used in Elastic Search [1] )

Machine Learning (or ML) is an area of Artificial Intelligence (AI) that is a set of statistical techniques for problem solving.

Machine Learning by itself is a set of algorithms that is used to do better NLP, better vision, better robotics etc. It is not an AI field in itself, but a way to solve real AI problems.

Today ML is used for self-driving cars (vision research from Figure 10-2), fraud detection, price prediction, and even NLP.

In order to apply ML techniques to NLP problems, we need to usually convert the unstructured text into a structured format, i.e. **tabular** format.

Deep Learning (which includes Recurrent Neural Networks, Convolution Neural Networks and others) is a type of Machine Learning approach.

Deep Learning is an extension of Neural Networks—which is the closest imitation of how the human brains work using neurons. Mathematically it involves running data through a large network of neurons—each of which has an activation function—the neuron is activated if that threshold is reached—and that value is propagated through the network.

Deep Learning is used quite extensively for vision based classification (e.g. distinguishing images of airplanes from images of dogs).

Deep Learning can be used for NLP tasks as well. However it is important to note that Deep Learning is a broad term used for a series of algorithms and it is just another tool to solve core AI problems that are highlighted above.

The image (Figure 10-3) shows graphically how NLP is related ML and Deep Learning. Deep Learning is one of the techniques in the area of Machine Learning—there are several other techniques such as Regression, K-Means, and so on.

ML and NLP have some overlap, as Machine Learning as a tool is often used for NLP tasks. There are several other things that you need for NLP—NER (Named Entity Recognizer), POS Tagged (Parts Of Speech Tagger identifies nouns, verbs and other part of speech tags in text).

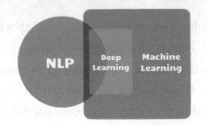

Figure 10-3　NLP has relationship with ML and Deep Learning

NLP has a strong linguistics component (not represented in the image), that requires an understanding of how we use language. The art of understanding language involves understanding humor, **sarcasm**, subconscious bias in text, etc.

Once we can understand that is means to be sarcastic (**yeah** right!) we can encode it into a machine learning algorithm to automatically discover similar patterns for us statistically.

To summarize, in order to do any NLP, you need to understand language. Language is different for different **genres** (research papers, blogs, twitter have different writing styles), so there is a strong need of looking at your data manually to get a feel of what it is trying to say to you, and how you—as a human would analyze it.

Once you figure out what you are doing as a human reasoning system (ignoring **hash tags**, using smiley faces to imply sentiment), you can use a relevant ML approach to automate that process and **scale** it.

 **Words**

highlight[ˈhailait] v. 强调, 突出
confluence[ˈkɔnfluəns] n. 汇合, 汇流点
insight[ˈinsait] n. 洞察力, 洞察
sentiment[ˈsentimənt] n. 感情, 情绪
internal[inˈtəːnl] n. 内部特征

tabular[ˈtæbjələ(r)] adj. 列成表格的
sarcasm[ˈsɑːkæzəm] n. 讽刺, 挖苦, 嘲笑
yeah[jeə] int. 是, 对
genre[ˈʒɔnrə] n. 类型, 种类
scale[skeil] v. 缩放, 测量

 **Phrases**

hash tag　　标签

 **Notes**

[1] Elastic Search 是一个基于 Lucene 的搜索服务器。它提供了一个分布式多用户能力的全文搜索引擎, 基于 RESTful Web 接口。Elastic Search 是用 Java 语言开发的, 并作为 Apache 许可条款下的开放源码发布, 是一种流行的企业级搜索引擎。Elastic Search 可

用于云计算中,能够实现实时搜索,稳定、可靠、快速,安装使用方便。官方客户端在 Java、.NET(C♯)、PHP、Python、Apache Groovy、Ruby 和许多其他语言中都是可用的。根据 DB-Engines 的排名,Elastic Search 是最受欢迎的企业搜索引擎,其次是 Apache Solr(也是基于 Lucene)。

  **Exercises**

I. Read the following statements carefully, and decide whether they are true (T) or false (F) according to the text.

　　____ 1. Artificial Intelligence (AI) is an area of Machine Learning (or ML).

　　____ 2. K-means is one of the techniques in the area of Machine Learning.

　　____ 3. Speech is highlighted in Figure 10-2.

　　____ 4. Artificial Intelligence is a part of the greater field of Computer Science.

　　____ 5. Deep Learning can be used for vision tasks as well as for NLP tasks.

II. Choose the best answer to each of the following questions according to the text.

　　1. Which of the following is not in Figure 10-2? (　　)

　　　　A. NLP

　　　　B. Vision

　　　　C. Robotics

　　　　D. Speech

　　2. Which of the following is an extension of Neural Networks? (　　)

　　　　A. Software Engineering

　　　　B. Deep Learning

　　　　C. Computer Science

　　　　D. None of the above

　　3. Which of the following is not right? (　　)

　　　　A. Artificial Intelligence (AI) is an area of Machine Learning (or ML).

　　　　B. Speech is highlighted in Figure 10-2.

　　　　C. Computer Science is a part of the greater field of Artificial Intelligence.

　　　　D. All of the above

III. Fill in the blanks with the words or phrases chosen from the box. Change the forms where necessary.

```
use    base    stand    source    know    relate
enable    sequential    connect    design
```

**BERT**

　　BERT is an open ____1____ machine learning framework for Natural Language

Processing (NLP). BERT is ___2___ to help computers understand the meaning of ambiguous language in text by using surrounding text to establish context. The BERT framework was pre-trained ___3___ text from Wikipedia and can be fine-tuned with question and answer datasets.

BERT, which ___4___ for Bidirectional Encoder Representations from Transformers, is based on Transformers, a deep learning model in which every output element is ___5___ to every input element, and the weightings between them are dynamically calculated ___6___ upon their connection. (In NLP, this process is called attention.)

Historically, language models could only read text input ___7___—either left-to-right or right-to-left—but couldn't do both at the same time. BERT is different because it is designed to read in both directions at once. This capability, ___8___ by the introduction of Transformers, is ___9___ as bidirectionality.

Using this bidirectional capability, BERT is pre-trained on two different, but ___10___, NLP tasks: Masked Language Modeling and Next Sentence Prediction.

**IV. Translate the following passage into Chinese.**

**Word2vec**

Word2vec is a two-layer neural net that processes text. Its input is a text corpus and its output is a set of vectors: feature vectors for words in that corpus. While Word2vec is not a deep neural network, it turns text into a numerical form that deep nets can understand. Deeplearning4j implements a distributed form of Word2vec for Java and Scala, which works on Spark with GPUs.

Word2vec's applications extend beyond parsing sentences in the wild. It can be applied just as well to genes, code, likes, playlists, social media graphs and other verbal or symbolic series in which patterns may be discerned.

# Part 2

# Simulated Writing: Writing for Employment (II)

接 169 页

**3. 策划简历**

简历是学历、工作经历、技能和成就汇总而成的一至两页的总结。就像求职信一样,要为每一个潜在的雇主定制一份简历。一份好的简历介绍了申请人的背景、突出水平和成就,这些都是最有可能吸引目标雇主的东西。尽管简历很短,只是结构化的文档,但它们需要精心的策划、编辑和校对。为了让简历能够吸引雇主从而保障拿到每一个面试的机会,最好准备几个不同版本的简历。表 10-2 总结了制作简历的注意事项。

表 10-2　制作简历的注意事项

| 简历元素 | 适合提到 | 尽量避免 |
| --- | --- | --- |
| 主标题 | • 包括完整的名字和完整的地址<br>• 使用简单且专业的格式来凸显名字<br>• 提供私人邮箱地址来显得更加专业<br>• 在找工作的时候定期查看信息和邮件 | • 遗漏联系方式<br>• 列出无效的联系信息<br>• 包含工作邮箱<br>• 使用华而不实或标新立异的格式<br>• 显示年龄、婚姻状况和薪资要求 |
| 职业目标和资历概述 | • 如果向某个职业做出了承诺就要包括一个职业目标<br>• 列出最重要的资历而不是一个目标 | • 目标和工作描述不符就放进来<br>• 将职业目标确定为初级职位，而小看了你的才能 |
| 教育经历总结 | • 标明与职位相关的具体课程<br>• 在信息的多少之间找到一个平衡点<br>• 如果课程平均绩点大于或等于3.0，就标注出来 | • 上了大学还列出高中的信息<br>• 把所学的所有课程都列出来 |
| 工作经历 | • 使用简单而准确的描述<br>• 量化成就<br>• 用列表将经历格式化 | • 列出全部的工作职责或活动<br>• 使用被动词汇或者强调个人主义（如一直提到我） |
| 技能、活动和荣誉 | • 重点强调所拥有的适合这个职位的技能<br>• 提供书面需求证明的具体细节<br>• 使用行动动词来描述技能和活动 | • 荣誉过少或很小还列出来<br>• 假设读者知道这个荣誉的意义，而没有提供一个简单的解释 |

1) 主标题

在简历的最开始，常常要标明姓名、地址和其他的联系信息，包括电话号码和电子邮件地址等。开通语音邮件、微信、QQ 等可以让招聘者或雇主更加方便地留言。

2) 职业目标和资历总结

如果给一个帖子或广告写回复，那么可以在主标题后面包括一个"目标"部分。我们的目标是给这个职位空缺定制一份简历，来展示你的职业目标与这个职位要求有多相符。比较接近的趋势是将目标和资历概述融合在一起。要想招聘经理从可能会收到的数百人的简历中区分出你的那一份，就需要一个包括三到八条目标性很强并能够显示你是这个职位的理想人选的列表。图 10-4 显示了一份包括目标部分的简历。

3) 教育经历总结

如果你刚毕业，那么接下来就可以总结你的教育程度。列出就读学校的名称和地点、主修课程或研究的领域、第二学位或者辅修课程，以及已经获得的所有学位的时间。当然如果你在课程中做得很好，还可包括课程平均绩点。

4) 工作经历

过去的工作经验和专业成就向招聘经理展现了你如何能够融入他们的组织。对于曾经

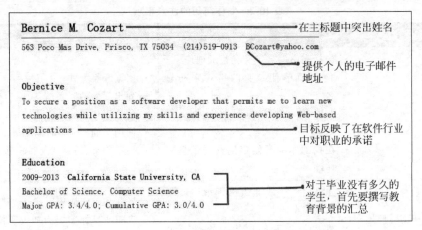

图 10-4 带有目标部分的简历

经历过的每个职位,要列出单位的名称和地点、职务和职位、聘用期(按年份,如 2016—2018 年;按月,如 2016 年 10 月—2018 年 3 月),以及一个简短的、对职责和成就的描述。

5) 技能、活动和荣誉

列出适合这个职位的特殊技能,如计算机水平、能讲的语言和行业认证等。也可以列出参加过的活动、加入的团体或曾获得的奖励,当然前提是有一些这样的经历。

### 4. 按时间顺序写简历

两种最流行的组织简历的方式是按时间展开和按功能展开。一份按时间顺序展开的简历,是按时间排序来呈现工作经历和教育经历,有时也被称为逆时间顺序的简历。从最近的工作经历或学习经历开始介绍,然后往前推进。尽量多包含近期的作品,少量谈及过去的经历。这种按时间顺序展开的风格是目前最受招聘人员和招聘经理欢迎的,因为它展现了你的就业经历的一个明确的时间线。如果曾有一个相对稳定的工作经历和几个相关的职位经历,那么按时间顺序排列的格式通常是你最好的选择。表 10-3 概括了按瞬间顺序写简历时该做的和不该做的注意事项。

表 10-3 按时间顺序写简历的注意事项

| 简历元素 | 适合提到 | 尽量避免 |
|---|---|---|
| 工作或教育经历 | • 如果是一个刚毕业的大学生,就先列出教育经历<br>• 如果工作业绩很显著,就先列出工作经历 | • 工作经验很多,全部列出来 |
| 格式 | • 按照大纲的格式来,用列表的方式列出细节<br>• 尽量保证顺序清晰且阅读体验好 | • 列出一大段文字<br>• 使用未经修改的标准模板,这对招聘人员来说太熟悉了 |
| 标题 | • 每个部分都以标题开头<br>• 使用标准的文本,例如资历、经验和教育的总结 | • 使用带有幽默口吻或者另类的标题<br>• 在校对的时候对标题一带而过 |

续表

| 简历元素 | 适合提到 | 尽量避免 |
| --- | --- | --- |
| 写作 | • 使用简洁但准确的语言<br>• 要检查拼写和语法错误<br>• 描述性语言以动词开头 | • 包括含糊不清或冗长的描述<br>• 有任何一个错字或者错误<br>• 使用被动动词 |
| 长度 | • 如果经历有限,简历只需一页<br>• 如果经历很丰富,或者想要申请的职位不止一个,那么就将简历拓展至两页 | • 让简历过长或过短<br>• 为了让简历只有一页或者有两页而刻意改变两边的界限 |

1）确定是先列出教育经历还是工作经验

如果你现在是一名学生或是一个刚刚毕业没有太多工作经历的大学生，那么就先列出你的教育经历。如果已经经历了三个或以上与你目前的工作前景有关的工作职位，那么对读者来说你的工作经历会更加有趣。随着工作经历的增多，你应该适当减少教育经历部分的内容，用一个简单的列表列出上学的学校和获得的学位。

2）格式化简历使其更具可读性

通常情况下，招聘人员和经理会快速扫视简历来寻找重要亮点，并跳过难以阅读的简历。用一两行总结工作经历和教育程度，并用（加在文字下面表示强调的）着重号来呈现工作和成就（参见图10-5）。使用标题、行、空格（没有文字或图形的区域）、颜色、粗体和斜体字体来布局页面，将注意力集中到具有清晰整洁的格式的部分。

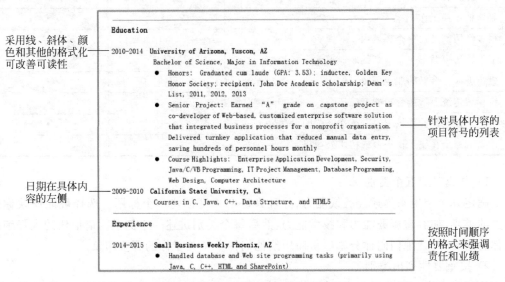

图 10-5　按照时间顺序撰写的简历

3）标题左对齐

将标题左对齐（不是中心）来组织页面，这缩短了每行的长度，使文字更易于阅读。左边统一的留白会将读者的目光吸引到页面上，只在标题做短暂的停留，这也预示着新章节的开始。

4）用明确有效的文字来写

认真编辑你的简历，使每一个句子都简短、准确并且积极向上。用一些积极向上的动词来展开你的句子，如明确了（clarified）、改进了（improved）、解决了（resolved）等，而不是做了（did）、是（was）、具有（had）等。用现在时态的动词来描述你目前的工作，用过去时态描述你过去的工作。

5）限定长度

如果工作经历少于三个职位，那么就将简历限定到一页。招聘人员自称他们更喜欢这种简短的格式，因为这样他们可以快速概览多份简历。当然，随着工作经验的增长，简历也可以扩展到两页。

### 5．书写功能型简历

功能型简历也称为技能简历，突出的是技术和能力，而不是工作经历。功能型简历由几个部分组成，分别展示才华、责任感和在大学课程中学习到的技能，还包括志愿者工作、实习、爱好或者其他方面。书写功能型简历适用于工作经历有限的学生、经常跳槽或者没有固定工作经历的人。功能简历也在某些行业很流行，如艺术、平面设计和网页制作。表10-4对时间顺序型和功能型简历进行了比较。除了写一份修改版时间顺序简历，还可以写一份功能型的简历来突出你的技能。

表 10-4 选择按时间顺序或功能型的方案

| 方　　案 | 按时间顺序 | 功　能　型 |
| --- | --- | --- |
| 工作经历有很多的空白 |  | • |
| 频繁地更换工作 |  | • |
| 很多经历都是在一个领域内 | • |  |
| 工作经历遵循了一个清晰的职业道路 | • |  |
| 想要进入一个新的职业领域 |  | • |
| 新领域传统上不接受功能型简历 | • |  |
| 是一个在选择的领域中没有工作经验的大学生 |  | • |
| 计划给一个在线工作平台或招聘公司提交简历 | • |  |

1）分成三到五个类别

确定几个类别来描述经验或能力，并且为每个类别创建一个标题。选择最能反映你的能力的类别，如客户服务能力和技术能力，然后每个类别讲述一个例子。请信任的人仔细阅读你的简历，并让他们向你分享这份简历给他们留下了关于你的哪些印象。

2）按重要性列出类别

按照对未来上司的重要性，给功能类别排序，组织各类别中的项目符号，先展示最相关的技术和能力。为了形成为职位定制的简历，要改变类别的顺序以满足每个职位。

3）用动词作为每个条目的开始

在每一个列表中的条目要简短、精炼，并且让人印象深刻。用行为动词作为每一个条目的开始，使它们更有趣，更易于阅读。在互联网上搜索"简历行为动词"（resume action verbs）来寻找可以使用的例子。

4) 提供工作经验简介

尽管你试图在技术和能力的基础上推销自己，招聘人员和招聘经理还是想知道你在哪里工作过。不推荐省略与工作经历相关的证明人。相反，应该在简历的最后总结性地简要列出工作经历，并标出每次工作经历的用人单位、时间段和职位，如图 10-6 所示。

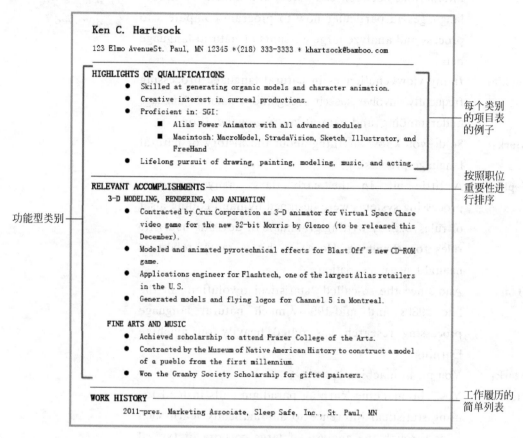

图 10-6　功能型简历的工作大纲

# Part 3

# Listening & Speaking

## Dialogue：Natural Language Processing

（As Mark is curious about natural language processing，he would like to discuss it with Sophie and Henry）

Mark： I think natural language processing is like computer vision，which is one of the most important areas of artificial intelligence.

Henry: Yes, I agree. Natural Language Processing (NLP) is a subfield of computer science, information engineering, and artificial intelligence concerned with [1] the interactions between computers and human (natural) languages, in particular how to program computers to process and analyze large amounts of natural language data.

[1] Replace with:
1. related to
2. involved

Sophie: In my view, challenges in natural language processing frequently involve speech recognition, natural language understanding, and natural language generation.

Mark: So do you know anything about the history of natural language processing?

Sophie: A little bit. In the early days, many language-processing systems were designed by hand-coding a set of rules, e.g. by writing grammars or devising heuristic rules for **stemming**. However, this is rarely robust to natural language variation.

Henry: And since the so-called "statistical revolution" in the late 1980s and mid-1990s, much natural language processing research has relied heavily on machine learning.

Mark: You mean machine learning?

Henry: Yes. The machine-learning paradigm calls instead for using statistical inference to automatically learn such rules through the analysis of large corpora of typical real-world. Example is a set of documents, possibly with human or computer **annotations**.

Mark: I guess there must be some machine-learning algorithms that have been applied to natural-language-processing tasks.

Sophie: You are right. Many different classes of machine-learning algorithms have been applied to natural-language-processing tasks. These algorithms take as input a large set of "features" that are generated from the input data. Some of the earliest-used algorithms, such as decision trees, produced systems of **hard** if-then rules similar to the systems of hand-written rules that were then common.

# Unit 10  Natural Language Processing

Henry: Increasingly, however, research has focused on statistical models, which make soft, probabilistic decisions based on attaching real-valued weights to each input feature. Such models have the advantage that they can express the relative certainty of many different possible answers rather than only one, producing more reliable results when such a model is included as a component of a larger system.

Mark: I heard systems based on machine-learning algorithms have many advantages over hand-produced rules?

Sophie: Absolutely. The learning procedures used during machine learning automatically focus on [2] the most common cases, whereas when writing rules by hand it is often not at all obvious where the effort should be directed.

> [2] Replace with:
> 1. emphasize on
> 2. concentrate on

Henry: And also automatic learning procedures can make use of statistical-inference algorithms to produce models that are robust to unfamiliar input (e.g. containing words or structures that have not been seen before) and to erroneous input (e.g. with misspelled words or words accidentally omitted). Generally, handling such input gracefully with hand-written rules—or, more generally, creating systems of hand-written rules that make soft decisions—is extremely difficult, **error-prone** and time-consuming.

Mark: Sounds interesting.

Sophie: Moreover, systems based on automatically learning the rules can be made more accurate simply by supplying more input data. However, systems based on hand-written rules can only be made more accurate by increasing the complexity of the rules, which is a much more difficult task.

Henry: In particular, there is a limit to the complexity of systems based on hand-crafted rules, beyond which the systems become more and more unmanageable. However, creating more data to input to machine-learning systems simply requires a corresponding increase in the number of man-hours worked, generally without significant increases in the complexity of the annotation process.

**Mark:** So much knowledge needs to know.

**Sophie & Henry:** That's right.

 **Exercises**

Work in a group, and make up a similar conversation by replacing the statements with other expressions on the right side.

 **Words**

stemming ['stemiŋ] *n.* 词干提取
annotation [ˌænə'teiʃn] *n.* 注释，注解
hard [hɑːd] *adj.* 难懂的，难以回答的
error-prone. 易于出错的

## Listening Comprehension: Speech Recognition

*Listen to the article and answer the following 3 questions based on it. After you hear a question, there will be a break of 15 seconds. During the break, you will decide which one is the best answer among the four choices marked (A),(B),(C) and (D).*

**Questions**

1. What is speech recognition also known as? ( )

   (A) Automatic speech recognition

   (B) Computer speech recognition

   (C) Speech to text

   (D) All of the above

2. What are systems that use training called? ( )

   (A) "Speaker independent" systems

   (B) "Speaker dependent" systems

   (C) "Text independent" systems

   (D) "Text dependent" systems

3. Which of the following is not right? ( )

   (A) Voice recognition refers to identifying the speaker.

   (B) Speaker identification refers to identifying the speaker.

   (C) Speaker identification refers to what speakers are saying.

   (D) All of the above

Unit 10　Natural Language Processing

 **Words**

enrollment[in'rəulmənt] n. 登记, 入伍
demotic[di'mɔtik] adj. 通俗的, 大众的
podcast['pɔdkɑ:st] n. 播客
radiology[ˌreidi'ɔlədʒi] n. 影像诊断学, 放射学

term[tɜ:m] v. 把……叫做
authenticate[ɔ:'θentikeit] v. 证实, 鉴别, 验证
evidence['evidəns] v. 证明

 **Phrases**

call routing　呼叫路由选择
collect call　对方付费电话, 拨打对方付费的电话

## Dictation：Natural Language Understanding

This article will be played three times. Listen carefully, and fill in the numbered spaces with the appropriate words you have heard.

Natural-Language Understanding (NLU) or Natural-Language ___1___ (NLI) is a ___2___ of natural-language processing in artificial intelligence that ___3___ with machine reading ___4___. Natural-language understanding is ___5___ an AI-hard problem.

There is ___6___ commercial interest in the field because of its application to ___7___ reasoning, machine translation, question answering, news-gathering, text ___8___, voice-activation, ___9___, and large-scale content analysis.

NLU is the post-processing of text, after the use of NLP ___10___ (identifying parts-of-speech, etc.), that ___11___ context from recognition devices (Automatic Speech Recognition [ASR], vision recognition, last conversation, ___12___ words from ASR, personalized ___13___, microphone **proximity** etc.), in all of its forms, to ___14___ meaning of ___15___ and **run-on** sentences to execute an intent from typically voice commands. A NLU has an **ontology** around particular product **vertical** that is used to figure out the ___16___ of an intent. A NLU has a defined list of ___17___ intents that ___18___ the message **payload** from designated contextual information recognition sources. The NLU will provide back multiple message outputs to separate services (software) or resources (hardware) from a single derived intent (response to voice command initiator with visual sentence (___19___ or spoken) and transformed voice command message too different output messages to be ___20___ for M2M communications and actions).

## Words

proximity[prɒk'simətɪ] n. 接近，邻近
run-on  词意连贯的，连续接排的
ontology[ɒn'tɒlədʒɪ] n. 本体，本体论

vertical['vɜːtɪkl] n. 垂直面，垂直位置
payload['peɪləʊd] n. 有效负载

# Glossary

## A

| | | | |
|---|---|---|---|
| a cascade of | | | 一系列，一连串 |
| a priori | | | 推理的 |
| a set of | | | 一组 |
| absorption | [əbˈzɔːpʃn] | n. | （液体、气体等的）吸收 |
| abstract | [ˈæbstrækt] | v. | 提取 |
| ace | [eis] | n. | 幺点 |
| act upon | | | 作用于 |
| actuator | [ˈæktjueitə] | n. | 效应器 |
| acyclic | [ˌeiˈsaiklik] | adj. | 非循环的 |
| acyclic graph | | | 非循环图 |
| address | [əˈdres] | v. | 处理(问题) |
| add up to | | | 合计达，总计达 |
| adjacent to | | | 临近的，靠近的 |
| adorable | [əˈdɔːrəbl] | adj. | 可爱的，值得敬重的 |
| advance | [ədˈvɑːns] | v. | 提出 |
| affine transformation | | | 仿射变换 |
| afresh | [əˈfreʃ] | adv. | 重新，再度 |
| aggregate | [ˈæɡriɡət i] | v. | 合计 |
| aggregate | [ˈæɡriɡət] | adj. | 聚合的，集合的 |
| akin to | | | 类似于，同类 |
| align with | | | 使一致，对准 |
| all about | | | 到处，各处 |
| among others | | | 除了别的之外，尤其 |
| amplitude | [ˈæmplitjuːd] | n. | 振幅，幅度 |
| an ocean of | | | 极多的，无穷无尽的 |
| and so | | | 因此，所以 |
| annotation | [ˌænəˈteiʃn] | n. | 注释，注解 |
| approach | [əˈprəutʃ] | v. | 处理 |
| approximation | [əˌprɔksiˈmeiʃn] | n. | 近似法 |
| archive | [ˈɑːkaiv] | v. | 把……存档 |
| argue | [ˈɑːɡjuː] | v. | 证明，说服 |
| argument | [ˈɑːɡjumənt] | n. | 论证，论据 |
| arise | [əˈraiz] | v. | 出现，上升 |

| | | | |
|---|---|---|---|
| arm race | | | 军备竞赛 |
| around | [əˈraund] | adv. | 到处,大约 |
| array | [əˈrei] | n. | 数组 |
| article | [ˈɑːtikl] | n. | 冠词 |
| at a distance | | | 在远处,有相当距离 |
| at first glance | | | 乍一看,初看 |
| at hand | | | 在手边,在附近 |
| attribute | [əˈtribjuːt, ˈætribjuːt] | n. | 属性 |
| augment | [ɔːgˈment] | v. | 增加,增大 |
| authentic | [ɔːˈθentik] | adj. | 真正的,真实的,可信的 |
| authenticate | [ɔːˈθentikeit] | v. | 证实,鉴别,验证 |
| average people | | | 普通人,一般人 |
| awkward | [ˈɔːkwəd] | adj. | 尴尬的,不合适的 |
| axiom | [ˈæksiəm] | n. | 公理 |
| axis | [ˈæksis] | n. | 轴,轴线 |
| axon | [ˈæksən] | n. | 轴突 |

## B

| | | | |
|---|---|---|---|
| backgammon | [ˈbækgæmən; ˌbækˈgæmən] | n. | 西洋双陆棋戏 |
| backpropagation | | | 反向传播 |
| backup | [ˈbækʌp] | n. | 备份,支持 |
| be adept at | | | 擅长,精湛纯熟 |
| be concerned with | | | 涉及 |
| be of | | | 具有……性质 |
| be opposed to | | | 与……相对,反对…… |
| be thought of as | | | 把……看作,被认为是 |
| beet | [biːt] | n. | 甜菜 |
| benchmark | [ˈbentʃmɑːk] | n. | 基准 |
| benevolent | [bəˈnevələnt] | adj. | 仁慈的,慈善的 |
| betray | [biˈtrei] | v. | 背叛(原则或信仰) |
| bias | [ˈbaiəs] | n. | 偏差 |
| binocular stereopsis | | | 双目立体视觉 |
| bioinformatics | [ˌbaiəuinfəˈmætiks] | n. | 生物信息学 |
| blank wall | | | 无法克服的障碍 |
| blunt | [blʌnt] | adj. | 生硬的,直率的 |
| blurry | [ˈblɜːri] | adj. | 模糊的,不清楚的 |
| board game | | | 棋盘游戏 |
| bot | [bɔt] | n. | 网上机器人,自动程序 |
| bounded rationality | | | 有限理性 |

| | | | |
|---|---|---|---|
| break down into | | | 分解成…… |
| bulk | [bʌlk] | n. | 大多数,大部分 |
| buzz | [bʌz] | n. | 时髦的(词语、想法或活动) |
| by convention | | | 按照惯例 |

## C

| | | | |
|---|---|---|---|
| call routing | | | 呼叫路由选择 |
| caption | [ˈkæpʃn] | n. | 标题 |
| casualty | [ˈkæʒuəlti] | n. | (战争或事故的)伤员,遇难者,受害者 |
| centroid | [ˈsentrɔid] | n. | 质心,形心 |
| characterize | [ˈkærəktəraiz] | v. | 特征化,表示,表现 |
| check | [tʃek] | n. | 支票 |
| checker | [ˈtʃekə(r)] | n. | 国际跳棋 |
| chemical composition | | | 化学组成(成分) |
| chore | [tʃɔː(r)] | n. | 家庭杂务 |
| claim | [kleim] | n. | (向公司等)索赔 |
| cluster | [ˈklʌstə(r)] | n. | 群集,簇,集群 |
| clustering | [ˈklʌstəriŋ] | n. | 聚类 |
| coin | [kɔin] | v. | 创造(新词,短语),杜撰 |
| collect call | | | 对方付费电话,拨打对方付费的电话 |
| come across | | | 偶然遇见 |
| come down with | | | 染上病 |
| come into play | | | (使)开始起作用 |
| come to the rescue | | | 救援,前来营救 |
| come up with | | | 提出,想出 |
| commonality | [kɔməˈnæliti] | n. | 公共,共性 |
| commonsense | [ˈkɔmənˈsens] | adj. | 具有常识的,常识的 |
| commutativity | [kə,mjuːtəˈtiviti] | n. | 交换性 |
| complement | [ˈkɔmplim(ə)nt] | v. | 补足,补助 |
| confluence | [ˈkɔnfluəns] | n. | 汇合、汇流点 |
| conjecture | [kənˈdʒektʃə(r)] | n. | 推测,猜想 |
| conjunction | [kənˈdʒʌŋkʃn] | n. | 连词,结合 |
| connotation | [ˌkɔnəˈteiʃn] | n. | 内涵 |
| contextual | [kənˈtekstʃuəl] | adj. | 上下文的 |
| convention | [kənˈvenʃn] | n. | 习俗,常规,惯例 |
| convergence | [kənˈvɜːdʒəns] | n. | 趋同,融合,一体化 |
| converse | [kənˈvɜːs; ˈkɔn-] | v. | 交谈,谈话 |

| | | | |
|---|---|---|---|
| convolutional | [ˌkɔnvəˈluːʃn(ə)l] | *adj.* | 卷积的 |
| cortex | [ˈkɔːteks] | *n.* | 皮质,皮层 |
| cortical | [ˈkɔːtikəl] | *adj.* | 皮层的,皮质的 |
| couple | [ˈkʌpl] | *v.* | 结合,连接 |
| cover | [ˈkʌvə(r)] | *v.* | 行走(一段路程) |
| credit assignment path | | | 权重分配路径 |
| creditworthiness | [ˈkreditwɜːrðinəs] | *n.* | 信誉,信用可靠程度 |
| crossover | [ˈkrɔsəuvə(r)] | *n.* | 组合交叉 |
| curvature | [ˈkɜːvətʃə(r)] | *n.* | 弯曲,[数]曲率 |
| cut down on | | | 减少,节省 |

## D

| | | | |
|---|---|---|---|
| decide on | | | 选定,决定 |
| deck | [dek] | *n.* | 一副扑克牌 |
| declarative | [diˈklærətiv] | *adj.* | 说明的,陈述的 |
| dedicated to | | | 致力于 |
| deductive | [diˈdʌktiv] | *adj.* | 演绎的,推论的,推断的 |
| demotic | [diˈmɔtik] | *adj.* | 通俗的,大众的 |
| dendrite | [ˈdendrait] | *n.* | 树突 |
| denomination | [diˌnɔmiˈneiʃn] | *n.* | 面额 |
| denote | [diˈnəut] | *v.* | 表示,指示 |
| designate | [ˈdezigneit] | *v.* | 指定,指派 |
| discrete | [diˈskriːt] | *adj.* | 离散的,不连续的 |
| discriminative | [disˈkriminətiv] | *adj.* | 区分的,区别的 |
| disjunction | [disˈdʒʌŋkʃn] | *n.* | 析取,分离 |
| dismay | [disˈmei] | *n.* | 沮丧,灰心 |
| displacement | [disˈpleismənt] | *n.* | 位移 |
| dominate | [ˈdɔmineit] | *v.* | 支配,控制 |
| down-sample | | | 降低取样 |
| draw from | | | 从……中得到,从……提取 |

## E

| | | | |
|---|---|---|---|
| echo | [ˈekəu] | *v.* | 重复,附和(想法或看法) |
| eerie | [ˈiəri] | *adj.* | 可怕的,怪异的 |
| emphatically | [imˈfætikli] | *adv.* | 着重地,断然地 |
| empirically | [imˈpirikli] | *adv.* | 以经验为主地 |
| encompass | [nˈkʌmpəs] | *v.* | 包含,包括,涉及(大量事物) |
| end up with | | | 以……告终,以……结束 |
| endeavor in | | | 在……的努力 |

| | | | |
|---|---|---|---|
| enrollment | [in'rəulmənt] | n. | 登记,入伍 |
| entice | [in'tais] | v. | 引诱,诱惑,吸引 |
| entirety | [in'taiərəti] | n. | 全部,完全 |
| entry | ['entri] | n. | 条目 |
| equilibrium | [ˌiːkwi'libriəm; ˌekwi'libriəm] | n. | 平衡,均衡 |
| eradicate | [i'rædikeit] | v. | 根除,消灭,杜绝 |
| error-prone. | | | 易于出错的 |
| et cetera | | | 等等 |
| Euclidean | [juː'klidiən] | adj. | 欧几里得几何学的,欧几里得的 |
| evidence | ['evidəns] | v. | 证明 |
| excel at | | | 擅长于,擅长 |
| excitation | [ˌeksai'teiʃ(ə)n] | n. | 励磁,激发 |
| exclusive | [ik'skluːsiv] | adj. | 独有的,排外的,专一的 |
| exhaustive | [ig'zɔːstiv] | adj. | 彻底的,穷尽的 |
| explicit | [ik'splisit] | adj. | 明确的,明白的 |
| exponentially | [ˌekspə'nenʃ(ə)li] | adv. | 以指数方式 |

## F

| | | | |
|---|---|---|---|
| factorial | [fæk'tɔːriəl] | n. | 阶乘 |
| fade | [feid] | adj. | 平淡的,乏味的 |
| fallacy | ['fæləsi] | n. | 谬论,谬误 |
| fatigue | [fə'tiːg] | n. | 疲劳,疲乏 |
| featureless | ['fiːtʃələs] | adj. | 一般的,平淡无奇的 |
| featurization | [fiːtʃərai'zeiʃən] | n. | 特征化,特性化 |
| feedforward | ['fiːdfɔːwəd] | n. | 前馈(控制) |
| fibrous | ['faibrəs] | adj. | 纤维的 |
| figure out | | | 想出,理解,解决 |
| file | [fail] | v. | 把……归档 |
| fine-tune | | | 调整 |
| fitness | ['fitnəs] | n. | 适应度,健康 |
| flat | [flæt] | n. | 少许电量 |
| flattened matrix | | | 扁平矩阵 |
| flood | [flʌd] | n. | 一大批 |
| folklore | ['fəuklɔː(r)] | n. | 民间传说 |
| follow the lead | | | 效仿,照样行事 |
| formal language | | | 形式语言 |
| frame | [freim] | n. | 框架 |
| frame | [freim] | n. | 帧,画面 |

| | | | |
|---|---|---|---|
| freak out | | | 崩溃,使处于极度兴奋中 |
| freeway | [ˈfriːwei] | n. | 高速公路 |
| fruitful | [ˈfruːtfl] | adj. | 富有成效的 |
| function | [ˈfʌŋkʃn] | n. | 函数 |
| fusion | [ˈfjuːʒn] | n. | 融合,熔化 |

## G

| | | | |
|---|---|---|---|
| gated recurrent unit | | | 门控循环单元 |
| gene | [dʒiːn] | n. | 基因 |
| genre | [ˈʒɒnrə] | n. | 类型,种类 |
| GO | [gəu] | n. | 围棋 |
| gradient | [ˈgreidiənt] | n. | 梯度,渐变 |
| gradient descent | | | 梯度下降 |
| grayscale | [ˈgreiskeil] | n. | 灰度 |
| grocer | [ˈgrəusə(r)] | n. | 杂货店,食品商 |
| gut | [gʌt] | adj. | 本能的,直觉的,简单的 |

## H

| | | | |
|---|---|---|---|
| hand-crafted | | | 手工制作的 |
| hand-engineered | | | 人工提取的 |
| hand-tuning | | | 手动调谐 |
| handy | [ˈhændi] | adj. | 有用的,便利的 |
| hard | [hɑːd] | adj. | 难懂的,难以回答的 |
| hash tag | | | 标签 |
| head start | | | 抢先起步的优势,有利的开端 |
| heavily | [ˈhevili] | adv. | 在很大程度上 |
| hibiscus | [hiˈbiskəs; haiˈbiskəs] | n. | 木槿,芙蓉花 |
| highlight | [ˈhailait] | v. | 强调,突出 |
| household | [ˈhaushəuld] | adj. | 家喻户晓的 |
| hype | [haip] | n. | 大肆宣传,炒作 |

## I

| | | | |
|---|---|---|---|
| identification | [ai͵dentifiˈkeiʃn] | n. | 识别,身份证明 |
| image | [ˈimidʒ] | n. | 生动的描绘,概念 |
| image captioning | | | 图像标注 |
| imagery | [ˈimidʒəri] | n. | 影像,意象,形象化 |
| implicit | [imˈplisit] | adj. | 含蓄的,暗中的 |
| improve at | | | 提升,改善 |

| | | | |
|---|---|---|---|
| in a nutshell | | | 简而言之,概括地说 |
| in between | | | 在中间 |
| in conjunction with | | | 连同,与协力 |
| in essence | | | 本质上,其实 |
| in response | | | 作为回答 |
| in turn | | | 依次,反过来 |
| inadvertently | [ˌinəd'vɜːtntli] | adv. | 无意地,不经意地 |
| increment | ['iŋkrəmənt] | n. | 增量 |
| indiscriminately | [ˌindi'skriminətli] | adv. | 不加选择地,任意地 |
| indisputable | [ˌindi'spjuːtəbl] | adj. | 不容置疑的,无可争辩的 |
| inductive | [in'dʌktiv] | adj. | 归纳的 |
| inertial | [i'nɜːʃl] | adj. | 惯性的,不活泼的 |
| infeasible | [in'fiːzib(ə)l] | adj. | 不可行的,不可实行的 |
| informed search | | | 启发式搜索 |
| ingenuity | [ˌindʒə'njuːəti] | n. | 独创性,足智多谋 |
| innate | [i'neit] | adj. | 先天的,固有的,与生俱来的 |
| innocent-looking | | | 看上去无恶意的,看上去无害的 |
| inside of | | | 在……之内,少于 |
| insider trading | | | 内线交易 |
| insight | ['insait] | n. | 洞察力,洞察 |
| instance | ['instəns] | n. | 实例 |
| instantiation | [instænʃi'eiʃən] | n. | 实例化,例示 |
| instrument | ['instrəmənt] | v. | 给……装测量仪器 |
| intellect | ['intəlekt] | n. | 智力,才智 |
| internal | [in'tɜːnl] | n. | 内部特征 |
| interrogator | [in'terəgeitə(r)] | n. | 讯问者,询问机 |
| intrinsically | [in'trinzikli; in'trinsikli] | adj. | 内在地,本质地,固有地 |
| invariably | [in'veəriəbli] | adv. | 总是,不变地,一定地 |

## J

| | | | |
|---|---|---|---|
| justification | [ˌdʒʌstifi'keiʃn] | n. | 理由,认为有理 |

## K

| | | | |
|---|---|---|---|
| knock over | | | 打翻,撞倒 |

## L

| | | | |
|---|---|---|---|
| landmark | ['lændmɑːk] | adj. | 有重大意义或影响的 |
| latent | ['leitnt] | adj. | 潜在的,潜伏的 |

| | | | |
|---|---|---|---|
| lateral | [ˈlætərəl] | *adj.* | 横向的,侧面的 |
| lethal | [ˈli:θl] | *adj.* | 致命的,致死的 |
| leverage | [ˈli:vərɪdʒ] | *v.* | 利用 |
| literally | [ˈlɪtərəli] | *adv.* | 真正地,确实地,简直 |
| literature | [ˈlɪtrətʃə(r)] | *n.* | 文献 |
| living-being | | | 有机体,生物 |
| local optimum | | | 局部最优,局部优化 |
| longitudinal | [ˌlɒŋgɪˈtju:dɪnl; ˌlɒndʒɪˈtju:dɪnl] | *adj.* | 纵向的 |

# M

| | | | |
|---|---|---|---|
| make out | | | 理解,辨认出 |
| malevolent | [məˈlevələnt] | *adj.* | 恶毒的,有恶意的 |
| manifest | [ˈmænɪfest] | *v.* | 表明 |
| manifestation | [ˌmænɪfeˈsteɪʃn] | *n.* | 表现,显示 |
| map | [mæp] | *v.* | 映射 |
| mean | [mi:n] | *n.* | 平均数,平均值 |
| memory footprint | | | 内存占用 |
| meta-reasoning | | | 元推理 |
| metric | [ˈmetrɪk] | *n.* | 度量标准 |
| mimic | [ˈmɪmɪk] | *v.* | 模仿 |
| mitigate | [ˈmɪtɪgeɪt] | *v.* | 减轻,缓和,缓解 |
| modality | [məʊˈdæləti] | *n.* | 形式,形态 |
| momentum | [məˈmentəm] | *n.* | 动量,动力 |
| monotonic | [ˌmɒnə(ʊ)ˈtɒnɪk] | *adj.* | 单调的,无变化的 |
| more of | | | 更大程度上 |
| motor reaction | | | 动作反应,运动反应 |
| much like | | | 就像,很像 |
| multilayer perception | | | 多层感知器 |
| multiplier | [ˈmʌltɪplaɪə(r)] | *n.* | 乘数 |
| mutation | [mju:ˈteɪʃn] | *n.* | 变异,突变 |

# N

| | | | |
|---|---|---|---|
| narrow down | | | 缩小 |
| negation | [nɪˈgeɪʃn] | *n.* | 否定,否认 |
| neuron | [ˈnjʊərɒn] | *n.* | 神经元 |
| nomenclature | [nəˈmeŋklətʃə(r)] | *n.* | 术语,命名法 |
| nonsense | [ˈnɒnsns] | *adj.* | 荒谬的 |
| notable | [ˈnəʊtəbl] | *adj.* | 值得注意的,显著的 |
| notion | [ˈnəʊʃn] | *n.* | 概念,想法,意图 |

| | | | |
|---|---|---|---|
| novel | [ˈnɒvl] | adj. | 新奇的,异常的 |
| nuisance | [ˈnjuːsns] | n. | 损害,麻烦事 |

## O

| | | | |
|---|---|---|---|
| obedient | [əˈbiːdiənt] | adj. | 顺从的,服从的 |
| observation | [ˌɒbzəˈveiʃn] | n. | 数据点 |
| obviate | [ˈɒbvieit] | v. | 排除,避免 |
| odometry | [əˈdɒmitri] | n. | 量距,测程法 |
| offspring | [ˈɒfspriŋ] | n. | 后代,子孙 |
| on-board | | | 在船(或飞机、车)上的 |
| on the surface | | | 在表面上,外表上 |
| ontology | [ɒnˈtɒlədʒi] | n. | 本体,本体论 |
| optic nerve | | | 视神经 |
| order | [ˈɔːdə(r)] | v. | 整理 |
| out of malice | | | 出于恶意 |
| outperform | [ˌautpəˈfɔːm] | v. | (效益上)超过,胜过 |
| outset | [ˈautset] | n. | 开始,开端 |
| outsmart | [ˌautˈsmɑːt] | v. | 比……更聪明,用计谋打败 |
| oval | [ˈəuvl] | n. | 椭圆型 |
| over and over | | | 反复,再三 |
| over time | | | 随着时间的过去 |

## P

| | | | |
|---|---|---|---|
| pacemaker | [ˈpeismeikə(r)] | n. | 心脏起搏器 |
| paradigm | [ˈpærədaim] | n. | 范式,范例 |
| parameter | [pəˈræmitə(r)] | n. | 参数 |
| parameterized | [pəˈræmitəraizd] | adj. | 参数化的 |
| parrot | [ˈpærət] | n. | 鹦鹉 |
| parser | [ˈpɑːsə] | n. | 词法分析器 |
| patch | [pætʃ] | n. | 小块土地 |
| pathfinding | [ˈpæθˌfaindiŋ] | n. | 寻找目标,探险 |
| payload | [ˈpeiləud] | n. | 有效负载 |
| permute | [pəˈmjuːt] | v. | 改变……的次序,重新排列 |
| persuasive | [pəˈsweisiv] | adj. | 有说服力的 |
| philanthropist | [fiˈlænθrəpist] | n. | 慈善家,乐善好施的人 |
| phoneme | [ˈfəuniːm] | n. | 音素,音位 |
| phrase | [freiz] | n. | 词组,短语 |
| pick out | | | 挑选出 |
| pixel | [ˈpiksl] | n. | (显示器或电视机图像的) |

| | | | |
|---|---|---|---|
| | | | 像素 |
| plan out | | | 策划，为……做准备 |
| plane | [plein] | n. | 平面 |
| planetary | [ˈplænətri] | adj. | 行星的 |
| plausibl | [ˈplɔːzəbl] | adj. | 振振有词的，似乎合理的，似是而非的，似乎可信的 |
| podcast | [ˈpɔdkɑːst] | n. | 播客 |
| pooling layer | | | 池化层 |
| population | [ˌpɔpjuˈleiʃn] | n. | 种群 |
| pose | [pəuz] | v. | 造成，形成 |
| posterior | [pɔˈstiəriə(r)] | adj. | 其次的，较后的 |
| pouch | [pautʃ] | n. | 小袋 |
| pragmatic analysis | | | 语用分析 |
| predicate | [ˈpredikət，ˈpredikeit] | n. | 谓词 |
| prefer over | | | 更喜欢 |
| premise | [ˈpremis] | n. | 前提，假设 |
| prescribe | [priˈskraib] | v. | 规定 |
| prick | [prik] | n. | 刺 |
| probability | [ˌprɔbəˈbiləti] | n. | 概率 |
| programmatic | [ˌprəugrəˈmætik] | adj. | 有计划的，按计划的 |
| prohibitively | [prəˈhibətivli] | adv. | 过高地，过分地 |
| project | [ˈprɔdʒekt] | v. | 表现，设计 |
| property | [ˈprɔpəti] | n. | 财产，所有权 |
| proposition | [ˌprɔpəˈziʃn] | n. | 命题 |
| protein | [ˈprəutiːn] | n. | 蛋白质 |
| proximity | [prɔkˈsiməti] | n. | 接近，邻近 |
| prudent | [ˈpruːdnt] | adj. | 谨慎的，慎重的 |

## Q

| | | | |
|---|---|---|---|
| quest | [kwest] | n. | 追求，寻找 |

## R

| | | | |
|---|---|---|---|
| radiologist | [ˌreidiˈɔlədʒist] | n. | 放射科医生，放射线研究者 |
| radiology | [ˌreidiˈɔlədʒi] | n. | 影像诊断学，放射学 |
| rating | [ˈreitiŋ] | n. | 等级评定，等级 |
| receptive field | | | 感受野 |
| reciprocate | [riˈsiprəkeit] | v. | 互换，报答 |
| recombination | [riːˌkɔmbiˈneiʃ(ə)n；ˌriːkɔmb-] | n. | 重组 |
| recommender system | | | 推荐系统 |

| | | | |
|---|---|---|---|
| recurrent | [riˈkʌrənt] | adj. | 循环的 |
| recursive | [riˈkəːsiv] | adj. | 循环的 |
| reflex machine | | | 反射机 |
| refrain from | | | 忍住,抑制,制止 |
| regression | [riˈgreʃn] | n. | 回归 |
| regularity | [ˌregjuˈlærəti] | n. | 正则性,规律性,规则性 |
| reinforcement learning | | | 强化学习 |
| resonance | [ˈrezənəns] | n. | 共振 |
| retina | [ˈretinə] | n. | 视网膜 |
| retract | [riˈtrækt] | v. | 取消,收回,缩回 |
| RNA-seq | | | 转录组测序技术(RNA sequencing) |
| rollout | | | 首次展示 |
| rule of thumb | | | 经验法则 |
| run-on | | | 词意连贯的,连续接排的 |

## S

| | | | |
|---|---|---|---|
| sarcasm | [ˈsɑːkæzəm] | n. | 讽刺,挖苦,嘲笑 |
| scale | [skeil] | v. | 缩放,测量 |
| scary | [ˈskeəri] | adj. | (事物)可怕的,引起惊慌的 |
| schema | [ˈskiːmə] | n. | 模式,计划 |
| schematically | [skiːˈmætikli] | adv. | 示意性地,图解地,计划性地 |
| scheme | [skiːm] | n. | 计划 |
| semantic | [siˈmæntik] | adj. | 语义的 |
| sentiment | [ˈsentimənt] | n. | 感情,情绪 |
| serve | [səːv] | v. | 对……有用,可作……用 |
| serve as | | | 用作,充当 |
| set out to | | | 打算,着手 |
| shared-weight | | | 参数共享 |
| shift invariant | | | 移位不变性 |
| shipment | [ˈʃipmənt] | n. | 装载的货物 |
| side effect | | | (药物的)副作用,意外的连带后果 |
| simplistic | [simˈplistik] | adj. | 过分简单化的 |
| simulate-annealing | | | 模拟退火法 |
| slant | [slɑːnt] | n. | 倾斜 |
| smoothly | [ˈsmuːðli] | adv. | 平稳地,顺利地 |
| snippet | [ˈsnipit] | n. | 片段 |
| so forth | | | 等等 |

| | | | |
|---|---|---|---|
| so on and so forth | | | 等等 |
| sonar | [ˈsəʊnɑː(r)] | n. | 声呐 |
| sound | [saʊnd] | adj. | 健全的,(非正式)非常棒的 |
| spade suit | | | 黑桃花色 |
| spectral | [ˈspektrəl] | adj. | 光谱的 |
| spectrum | [ˈspektrəm] | n. | 范围,领域 |
| spontaneously | [spɒnˈteɪniəsli] | adv. | 自发地,自然地 |
| squared distance | | | 距离平方 |
| stand out | | | 突出 |
| stand up to | | | 经得起,抵抗 |
| start off | | | 开始 |
| starts out with | | | 从……入手,开始 |
| steer | [iˈrædɪkeɪt] | v. | 驾驶(船、汽车等) |
| stem | [stem] | n. | 干,茎 |
| stemming | [ˈstemɪŋ] | n. | 词干提取 |
| step on | | | 踩上……,踏上…… |
| stereoscopic | [ˌsteriəˈskɒpɪk] | adj. | 立体的,实体镜的 |
| stimuli | [ˈstɪmjʊlaɪ] | n. | 刺激,刺激物 |
| stochastic | [stəˈkæstɪk] | n. | 随机的,猜测的 |
| strategize | [ˈstrætɪdʒaɪz] | v. | 制定战略 |
| strawberry | [ˈstrɔːbəri] | n. | 草莓 |
| strikingly | [ˈstraɪkɪŋli] | adv. | 显著地,突出地 |
| stubborn | [ˈstʌbən] | adj. | 难处理的,顽固的 |
| stuck | [stʌk] | adj. | (因困难)无法继续的,停滞不前的 |
| stumble upon | | | 偶然发现 |
| subarea | [ˈsʌbɛərɪə] | n. | 分区 |
| subject to | | | 受制于 |
| subsequently | [ˈsʌbsɪkwəntli] | adv. | 随后,其后 |
| subservient | [səbˈsɜːvɪənt] | adj. | 有用的,有帮助的 |
| such that | | | 如此……以致 |
| sum up | | | 计算……的总数 |
| surefire | [ʃʊəˈfaɪə] | adj. | 保准不会有错的,一定成功的 |
| susceptible to | | | 易受……影响的,对……敏感的 |
| swarm | [swɔːm] | n. | 一大群 |
| synapse | [ˈsaɪnæps] | n. | 突触 |
| system | [ˈsɪstəm] | n. | 方法 |

# T

| | | | |
|---|---|---|---|
| tabular | [ˈtæbjələ(r)] | adj. | 列成表格的 |
| take advantage of | | | 利用 |
| take into account | | | 考虑,顾及 |
| tally up | | | 结算 |
| task | [tɑːsk] | v. | 派给某人(任务) |
| tasty | [ˈteisti] | adj. | 美味的 |
| tax | [tæks] | v. | 使……负重担 |
| tax return | | | 纳税申报单 |
| tell apart | | | 区分,分辨 |
| temporal | [ˈtempərəl] | adj. | 暂时的,时间的 |
| tentacle | [ˈtentəkl] | n. | 触角 |
| tentative | [ˈtentətiv] | adj. | 试验性的,暂定的 |
| term | [tɜːm] | v. | 把……叫作 |
| test bed | | | 试验台 |
| textual data | | | 文本数据 |
| the like | | | 类似的东西 |
| theorem | [ˈθiərəm] | n. | 定理,原理 |
| therapeutic | [ˌθerəˈpjuːtik] | adj. | 治疗的,医疗的 |
| tile | [tail] | v. | 平铺显示 |
| think of as | | | 把…看作 |
| threshold | [ˈθreʃhəuld] | n. | 阈值,临界值 |
| thwart | [θwɔːt] | v. | 挫败,反对 |
| tilt | [tilt] | v. | 翘起,倾斜 |
| to date | | | 至今,迄今为止 |
| to the fullest | | | 充分地,达到最大程度 |
| tough | [tʌf] | adj. | 坚强的,不屈不挠的 |
| trade-off | | | 权衡,取舍 |
| transcribe | [trænˈskraib] | v. | 改编,转录,抄写 |
| traversal | [trəˈvɜːs(ə)l] | n. | 遍历 |
| trick | [trik] | v. | 欺骗,哄骗 |
| trust in | | | 信任 |
| tune | [tjuːn] | v. | 调整,使一致 |
| turn out to be | | | 证明是,结果是,原来是 |
| turn out | | | 结果是,证明是 |
| turnip | [ˈtɜːnip] | n. | 萝卜 |
| twirl | [twɜːl] | v. | (使)旋转,转动 |
| twist | [twist] | n. | (故事或情况的)转折,转变 |
| typology | [taiˈpɔlədʒi] | n. | 分类法 |

## U

| | | | |
|---|---|---|---|
| uncharted territory | | | 未知的领域 |
| underline | [ˌʌndəˈlain] | v. | 强调 |
| underlying | [ˌʌndəˈlaiiŋ] | adj. | 根本的 |
| undertaking | [ˌʌndəˈteikiŋ] | n. | 任务,事业 |
| uninformed search | | | 盲目搜索 |
| universal approximator | | | 泛逼近器 |
| unroll | [ʌnˈrəul] | v. | 展开,显示 |
| unwittingly | [ʌnˈwitiŋli] | adv. | 不经意地 |
| up for grabs | | | 大家有份 |

## V

| | | | |
|---|---|---|---|
| venture | [ˈventʃə(r)] | v. | 敢于冒险 |
| vertical | [ˈvɜːtikl] | n. | 垂直面,垂直位置 |
| vertices | [ˈvɜːtisiːz] | n. | 顶点,至高点(vertex 的复数) |
| vestibulogram | | | 针对前庭眼球震颤评估的图形记录 |
| vicinity | [vəˈsinəti] | n. | 邻近,附近 |
| visual field | | | 视野 |
| vomit | [ˈvɔmit] | n. | 呕吐 |

## W

| | | | |
|---|---|---|---|
| weight | [weit] | v. & n. | 加权,权重 |
| weight | [weit] | n. | 参数 |
| well-defined | | | 定义明确的 |
| well-established | | | 得到确认的 |
| well-justified | | | 合理的 |
| well-suited | | | 便利的,适当的 |
| whereas | [ˌweərˈæz] | conj. | 然而,鉴于 |
| whisker | [ˈwiskə(r)] | n. | 胡须 |
| with respect to | | | 关于,至于 |
| wittingly | [ˈwitiŋli] | adv. | 有意地 |
| work horse | | | 主力,主要设备 |
| wreak havoc | | | 造成严重破坏,肆虐 |

## Y

| | | | |
|---|---|---|---|
| yeah | [jeə] | int. | 是,对 |
| zoom in on | | | 聚焦于,推近 |

# Abbreviations

## A

a.k.a.                                                       .k.a. 亦称，又名（also known as）

## G

| | | |
|---|---|---|
| GAN | Generative Adversarial Network | 生成对抗网络 |
| GPU | Graphics Processing Unit | 图形处理器 |

## M

| | | |
|---|---|---|
| MRI | Magnetic Resonance Imaging | 核磁共振成像 |

## N

| | | |
|---|---|---|
| NARS | Non-Axiomatic Reasoning System | 非公理推荐系统 |

# Answers

## Unit 1

### Part 1  Reading & Translating

**Section A**

I. 1. F    2. F    3. F    4. T    5. T

II. 1. B    2. C    3. D

III.
1. approached  2. However   3. instead  4. known    5. while
6. seems       7. outperform 8. research 9. applied  10. performance

IV. 掌上人工智能

智能手机应用中逐渐展现出了越来越多的人工智能技术。例如，谷歌研发了 Google Goggles，它是一个提供虚拟搜索引擎的智能手机应用。只要用智能手机的摄像头拍摄一本书、某一地标或某一标记，Goggles 就会执行图像处理、图像分析以及文本识别，然后启动 Web 搜索来识别对象。如果讲英语的你正处在法国，你可以拍摄一张地标、菜单或其他文本的照片，然后 Goggles 会将其翻译为英文。除了 Goggles 以外，谷歌正在积极地研究声音对声音的语言翻译，很快你就可以用英语对着手机说话，然后让手机将之用西班牙语、中文或其他语言翻译出来。随着不断以创新的方式使用 AI，智能手机无疑会越来越智能。

**Section B**

I. 1. F    2. F    3. F    4. F    5. T

II. 1. C    2. D    3. C

III.
1. programmed    2. However   3. debated        4. inherently  5. proponents
6. individually  7. resolving 8. characteristics 9. observe     10. exhibits

IV. 物理智能体

物理智能体(机器人)是一个用来完成各项任务的可编程系统。简单的机器人可以用在制造行业，从事一些日常的工作，如装配、焊接或油漆。有些组织使用移动机器人去做一些日常的分发工作，如分发邮件或明信片到不同的房间。移动机器人可以在水下探测石油。

人形机器人是一种自治的移动机器人，它模仿人类的行为。虽然人形机器人只在科幻小说中流行，但是要使这种机器人能合理地与周围环境交互并从环境里发生的事件中学习，这里面还有很多工作要做。

### Part 3  Listening & Speaking

**Listening Comprehension**

1.（C）    2.（D）    3.（B）

**Original**

**Thinking Machines**

Computers are amazing devices. They can draw complex three-dimensional images, process the payroll of an entire corporation, and determine whether the bridge you're building will **stand up to** the pressure of the traffic expected. Yet they may have trouble understanding a simple conversation and might not be able to distinguish between a table and a chair.

Certainly a computer can do some things better than a human can. For example, if you are given the task of adding 1000 four-digit numbers together using pencil and paper, you could do it. But the task would take you quite a long time, and you very likely would make errors while performing the calculations. A computer could perform the same calculation in a fraction of a second without error.

However, if you are asked to point out the cat in the picture which includes a cat, you could do it without hesitation. A computer, by contrast, would have difficulty making that **identification** and might very well get it wrong. Humans bring a great deal of knowledge and reasoning capability to these types of problems; we are still struggling with ways to perform human-like reasoning using a computer.

In our modern state of technology, computers are good at computation, but less **adept at** things that require intelligence. The field of Artificial Intelligence (AI) is the study of computer systems that attempt to model and apply the intelligence of the human mind.

## Dictation

| | | | | |
|---|---|---|---|---|
| 1. autonomous | 2. directs | 3. achieve | 4. considered | 5. abstract |
| 6. called | 7. implementations | 8. emphasize | 9. behavior | 10. borrowed |
| 11. rational | 12. related | 13. studied | 14. practical | 15. modeling |
| 16. closely | 17. tasks | 18. refer | 19. operator | 20. mining |

# Unit 2

## Part 1 Reading & Translating

### Section A

I. 1. F  2. F  3. T  4. T  5. T
II. 1. C  2. A  3. C
III.
1. broken  2. combine  3. infer  4. defines  5. arguments
6. predicates  7. relationship  8. refers  9. between  10. purpose

IV. 推理和逻辑

推理是把思想构造成有效论点的行为。这可能是你每天都要做的事。当你作决定时,你在推理并构思不同的想法,然后把这些想法变成为什么你应该作出某个选择和非其他选择的理由。构造论点时,该论点可能是有效的也可能是无效的。一个有效的论点是在逻辑或事实基础上综合的推理。

归纳推理和演绎推理是命题逻辑的两种形式。命题逻辑是逻辑学的一个分支,它研究如何连接和/或修改整个命题、语句或句子,以形成更复杂的命题、语句或句子。归纳推理和演绎推理都使用命题逻辑来组织基于事实和推理的有效论据。这两种推理都有前提和结论,每种类型的推理如何得出结论的过程是不同的。

### Section B

I. 1. F  2. F  3. F  4. T  5. T
II. 1. A  2. B  3. C
III.
1. refer  2. called  3. accepted  4. expressed  5. practice
6. set  7. opposite  8. likely  9. equivalent  10. though

IV. 语义网络

语义网络(Semantic network,简称语义网)是 Quillian 于 1967 年提出的一种以图形形式表示知识的网络。语义网络是一种用于命题信息的知识表示技术,有时也称为命题网络。在知识表示中,语义网络是二维的。在数学上,语义网络被定义为带有标签的有向图。语义网络由链接、节点和链接标签组成。在图中,语义网络节点被描述为椭圆、圆或矩形,以显示物理对象、情景或概念等对象。链接可用于表示对象之间的关系。特定关系由链接标签指定。知识组织的基本结构是由关系组成的。

## Part 3　Listening & Speaking

### Listening Comprehension

　　1. (B)　　　　　2. (D)　　　　　3. (C)

**Original**

**Logical Reasoning**

　　Logical reasoning is a **system** of forming conclusions based on **a set of premises** or information. Commonly, logical reasoning is **broken down into** two major types called **deductive** and **inductive** reasoning. While the principles of logic can be used to create a strong **argument** for or against a conclusion, the system has several vulnerabilities, including the potential for untrue premises, **fallacies**, and intentional distortion of reason.

　　To reach a conclusion using logical reasoning, evidence or facts must first be presented. For instance, if a grocer wants to know if he is selling more **beets** than **turnips**, he may gather evidence about the amount of the two vegetables in recent **shipments**, how many have been sold, and if any product loss has occurred due to theft or damage. If his premises show that he sold 52 turnips and 75 beets in the same month, with no loss due to theft or damage, he can logically conclude that he sells more beets than turnips based on the evidence.

　　The type of reasoning in the above example is known as deductive reasoning. This type of logic occurs when the premises **add up to** a single, **indisputable** conclusion. Given that the premises are accurate, deductive reasoning can prove an absolute truth or fact. Inductive logic, by contrast, uses premises to determine a highly probable, but not absolute, conclusion. While inductive logical reasoning can be far more complex to understand than deductive reasoning, it generally forms the **bulk** of most logic-based arguments.

### Dictation

| | | | | |
|---|---|---|---|---|
| 1. developed | 2. directed | 3. edges | 4. concepts | 5. exact |
| 6. related | 7. sets | 8. therefore | 9. animal | 10. horse |
| 11. shown | 12. define | 13. subclass | 14. instance | 15. attribute |
| 16. object | 17. inheritance | 18. fact | 19. present | 20. infer |

# Unit 3

## Part 1　Reading & Translating

### Section A

Ⅰ. 1. F　　　　2. F　　　　3. T　　　　4. T　　　　5. T

Ⅱ. 1. B　　　　2. D　　　　3. A

Ⅲ.

| | | | | |
|---|---|---|---|---|
| 1. introduced | 2. fuzzy | 3. typically | 4. refer | 5. kinds |
| 6. some | 7. particular | 8. aimed | 9. based | 10. propositional |

Ⅳ. 非单调逻辑

　　日常推理大多是非单调的,因为它涉及风险,即我们会在演绎不足的前提下得出结论。我们知道什么时候冒这个险是值得的,甚至是必要的(例如,在医学诊断中)。然而,我们也意识到这样的推论是"可废止的"——因为新的信息可能会破坏旧的结论。传统上,各种可废止的但非常成功的推论(归纳理论、皮尔斯的诱拐理论、最佳解释的推论等)引起了哲学家们的注意。最近,逻辑学家开始从形式的角度来探讨这一现象,并在哲学、逻辑和人工智能的界面上形成了一个庞大的理论体系。

## Section B

Ⅰ. 1. F          2. F          3. F          4. T          5. F
Ⅱ. 1. D          2. B          3. D
Ⅲ.
1. given         2. either     3. despite    4. know       5. False
6. sure          7. likely     8. being      9. otherwise  10. have

Ⅳ. 概率推理

概率推理是应用概率概念来表示知识不确定性的一种知识表示方法。在概率推理中,我们将概率论与逻辑相结合来处理不确定性。

我们在概率推理中使用概率,是因为它提供了一种方法来处理由某人的懒惰或无知造成的不确定性。

在现实世界中,有很多情况下,某些事情的确定性是不确定的,比如"今天会下雨""某个人在某些情况下的行为""两队或两名球员之间的比赛"。这些都是不确定的描述,我们可以假设它会发生,但并不确定,所以此时我们使用概率推理。

# Part 3   Listening & Speaking

## Listening Comprehension

1.（C）          2.（D）          3.（D）

**Original**

**Fuzzy Logic**

Fuzzy logic is an approach to computing based on "degrees of truth" rather than the usual "true or false" (1 or 0) Boolean logic on which the modern computer is based.

The idea of fuzzy logic was first **advanced** by Dr. Lotfi Zadeh of the University of California at Berkeley in the 1960s. Dr. Zadeh was working on the problem of computer understanding of natural language. Natural language (like most other activities in life and indeed the universe) is not easily translated into the absolute terms of 0 and 1. (Whether everything is ultimately describable in binary terms is a philosophical question worth pursuing, but in practice much data we might want to feed a computer is in some state **in between and so**, frequently, are the results of computing.) It may help to see fuzzy logic as the way reasoning really works and binary or Boolean logic is simply a special case of it.

Fuzzy logic includes 0 and 1 as extreme cases of truth (or "the state of matters" or "fact") but also includes the various states of truth in between so that, for example, the result of a comparison between two things could be not "tall" or "short" but ".38 of tallness."

Fuzzy logic seems closer to the way our brains work. We **aggregate** data and form a number of partial truths which we aggregate further into higher truths which in turn, when certain **thresholds** are exceeded, cause certain further results such as **motor reaction**. A similar kind of process is used in neural networks, expert systems and other artificial intelligence applications. Fuzzy logic is essential to the development of human-like capabilities for AI, sometimes referred to as artificial general intelligence: the representation of generalized human cognitive abilities in software so that, faced with an unfamiliar task, the AI system could find a solution.

## Dictation

1. graphical        2. represent        3. Given          4. algorithms       5. sequences
6. called           7. indicates        8. significant    9. currently        10. faults
11. construct       12. uncomfortable   13. based         14. sufficient      15. hypothesis
16. managed         17. failed          18. distrust      19. eliminated      20. hierarchy

# Unit 4

## Part 1　Reading & Translating

### Section A

Ⅰ.　1. F　　　2. T　　　3. F　　　4. T　　　5. T

II. 1. B　　　2. C　　　3. A

III.

1. approached　　2. based　　3. evolved　　4. appear　　5. others

6. outperform　　7. commonsense　　8. techniques　　9. desired　　10. performance

IV. 旅行商问题

一名推销员想到访 N 个城市(需要经过每个城市)。我们怎样排列到访顺序才能使推销员的行程最短？此处需要最小化的目标函数是行程的长度(所有城市之间的距离按指定顺序的总和)。

要解决这个问题,需要：

(1) 配置设置：城市从 1 到 N 的所有可能排列。我们的目标是在这些排列中选择一个最佳排列。

(2) 重新安排策略：我们将遵循的策略是用随机路径替换部分路径,然后重新测试这个修改后的路径是否是最优的。

(3) 目标函数(这是最小化的目标)：某个特定顺序的所有城市之间距离的总和。

### Section B

Ⅰ.　1. F　　　2. T　　　3. T　　　4. T　　　5. T

Ⅱ.　1. D　　　2. B　　　3. A

III.

1. used　　2. generating　　3. longer　　4. chosen　　5. available

6. whether　　7. iteratively　　8. called　　9. wide　　10. turns

IV. 进化规划

当应用于程序开发时,使用遗传算法的方法称为进化规划(evolutionary programming)。此时,我们的目标是通过模拟进化过程开发程序,而不是直接编写程序。研究人员已经用函数式程序设计语言将进化规划技术应用于程序开发过程。该方法首先创建了一个包含各种函数的程序集合。初始集合中的函数构成了"基因池",而之后的各代程序将通过"基因池"来构建。接下来,我们允许进化过程执行很多代,期望每次通过上一代中的最佳组合生成新的一代,从而让目标问题的解决方案逐步进化。

## Part 3　Listening & Speaking

### Listening Comprehension

1. (D)　　　2. (A)　　　3. (D)

**Original**

**A* Search**

A* is a computer algorithm that is widely used in **pathfinding** and graph traversal. The algorithm efficiently plots a walkable path between multiple nodes, or points, on the graph.

On a map with many obstacles, pathfinding from points AA to BB can be difficult. A robot, for instance, without getting much other direction, will continue until it encounters an obstacle.

However, the A* algorithm introduces a heuristic into a regular graph-searching algorithm, essentially planning ahead at each step so a more optimal decision is made. With A*, a robot would instead find a path.

A* is an extension of Dijkstra's algorithm with some characteristics of Breadth-First Search (BFS).

Like Dijkstra, A* works by making a lowest-cost path tree from the start node to the target node.

What makes A* different and better for many searches is that for each node, A* uses a function f(n) that gives an estimate of the total cost of a path using that node. Therefore, A* is a heuristic function, which differs from an algorithm in that a heuristic is **more of** an estimate and is not necessarily provably correct.

## Dictation

| | | | |
|---|---|---|---|
| 1. algorithm | 2. expensive | 3. obviously | 4. speed | 5. options |
| 6. commonly | 7. optimal | 8. prior | 9. account | 10. goal |
| 11. node | 12. estimate | 13. direction | 14. lead | 15. note |
| 16. exploring | 17. reasonable | 18. scenarios | 19. efficient | 20. solve |

# Unit 5

## Part 1   Reading & Translating

### Section A

Ⅰ. 1. F    2. T    3. T    4. F    5. F
Ⅱ. 1. B    2. B    3. C
Ⅲ.
  1. commonly    2. variables    3. what    4. estimates    5. independent
  6. defined     7. dependent    8. many    9. called       10. can

Ⅳ. 支持向量机

支持向量机是一种有监督的学习算法，它将数据分为两类。它在最初训练时建立模型，并使用一系列已经分为两类的数据进行训练。支持向量机算法的任务是确定一个新的数据点应归入哪一类。这使得支持向量机成为一种非二元性的的线性分类器。

支持向量机算法不仅要将对象分类，而且要使图中的类别间的边界尽可能宽。

### Section B

Ⅰ. 1. F    2. T    3. T    4. F    5. T
Ⅱ. 1. A    2. D    3. D
Ⅲ.
  1. as          2. deal     3. used   4. understand   5. based
  6. reflects    7. method   8. like   9. advantage    10. labor

Ⅳ. 集成学习

可以通过训练许多集成学习工具以产生多种结果。单个的算法可以相互叠加，也可以依赖于一个系统评估多个方法的"模型桶"方法。在某些情况下，可以聚合或组合多个数据集。例如，地理研究项目可以使用多种方法来评估某种物质在地理空间中的分布。这类研究的一个问题是需要确保各种模型是独立的，并且数据的组合是实用的，还需要能在特定的情境中使用。

不同类型的统计软件包中都有集成学习方法。一些专家将集成学习描述为数据聚合的"众包"（crowd sourcing）。

## Part 3   Listening & Speaking

### Listening Comprehension

1.（D）    2.（D）    3.（A）

**Original**

**Supervised Learning**

Supervised learning is the machine learning task of learning a function that **maps** an input to an output based on example input-output pairs. It infers a function from labeled training data consisting of a set of training examples. In supervised learning, each example is a pair consisting of an input object (typically a vector) and a desired output value (also called the supervisory signal). A supervised learning algorithm analyzes the training data and produces an inferred function, which can be used for mapping new examples. An optimal scenario will allow for the algorithm to correctly determine the class labels for unseen instances. This requires the learning algorithm to generalize from the training data to unseen situations in a "reasonable" way.

When the training data contains explicit examples of what the correct output should be for given inputs, then we are within the supervised learning setting that we have covered so far. Consider the hand-written digit recognition problem. A reasonable data set for this problem is a collection of images of hand-written digits, and for each image, what the digit actually is. We thus have a set of examples of the form (image, digit).

The learning is supervised in the sense that some 'supervisor' has taken the trouble to look at each input, in this case an image, and determine the correct output, in this case one of the ten categories {0,1, 2,3,4,5,6,7,8,9}.

While we are on the subject of variations, there is more than one way that a data set can be presented to the learning process. Data sets are typically created and presented to us in their **entirety** at the **outset** of the learning process.

**Dictation**

1. branch      2. labeled         3. responding    4. absence      5. Alternatives
6. setting     7. classification  8. coins         9. clusters     10. unlabeled
11. identical  12. obvious        13. ambiguous    14. Nonetheless 15. viewed
16. patterns   17. categorize     18. general      19. various     20. properties

# Unit 6

## Part 1  Reading & Translating

### Section A

I. 1. F         2. T          3. F          4. T          5. F
II. 1. D        2. D          3. C
III.
1. corresponds    2. around    3. discussing    4. based    5. interchangeably
6. Various        7. distinguish    8. linear    9. non-linear    10. commonly

IV. 非线性激活函数

激活函数是定义神经元输出的任意函数。与神经网络中每个神经元相关联的激活函数根据该函数的输出来决定是否应该激活它。激活函数有三种类型：二元激活函数、线性激活函数和非线性激活函数。

神经网络的输入通常是线性变换（即输入×权重＋偏差），但大多数实际数据都是非线性的。因此，为了使输入非线性，使用了非线性激活函数。非线性激活是在网络中加入非线性的函数。

### Section B

I. 1. F         2. T          3. T          4. F          5. F
II. 1. D        2. C          3. D
III.
1. far         2. simplest    3. adjusting    4. Generally    5. corresponds
6. according   7. creating    8. Similarly    9. includes     10. noted

Ⅳ. 前馈神经网络

前馈神经网络作为神经网络设计的一个主要例子，其结构是有限的。信号从一个输入层传到另一层。一些前馈网络的结构甚至更简单。例如，单层感知器模型只有一层，前馈信号从一层移动到单个节点。具有更多层的多层感知器模型，也是前馈的。

自从科学家发明了第一个人工神经网络以来，科技界在建立更复杂的模型方面取得了各种进展：有递归神经网络和其他包含循环的设计，还有些模型涉及反向传播，反向传播模型中的机器学习系统本质上是通过系统发送回数据来优化的。前馈神经网络不涉及任何此类设计，因此它是一种独特的系统类型，有利于首次学习这些设计。

## Part 3　Listening & Speaking
### Listening Comprehension

1.（B）　　　　　2.（D）　　　　　3.（B）

**Original**

**Training Artificial Neural Networks**

An important feature of artificial neural networks is that they are not programmed in the traditional sense but instead are trained. That is, a programmer does not determine the values of the weights needed to solve a particular problem and then "plug" those values into the network. Instead, an artificial neural network learns the proper weight values via supervised training involving a repetitive process in which inputs from the training set are applied to the network and then the weights are adjusted by small **increments** so that the network's performance **approaches** the desired behavior.

Let us consider an example in which training an artificial neural network to solve a problem has been successful and perhaps more productive than trying to provide a solution by means of traditional programming techniques. The problem is one that might be faced by a robot when trying to understand its environment via the information it receives from its video camera. Suppose, for example, that the robot must distinguish between the walls of a room, which are white, and the floor, which is black. **At first glance**, this would appear to be an easy task: Simply classify the white pixels as part of a wall and the black pixels at part of the floor. However, as the robot looks in different directions or moves around in the room, various lighting conditions can cause the wall to appear gray in some cases whereas in other cases the floor may appear gray. Thus, the robot needs to learn to distinguish between walls and floor under a wide variety of lighting conditions. To accomplish this, we could build an artificial neural network whose inputs consist of values indicating the color characteristics of an individual pixel in the image as well as a value indicating the overall brightness of the entire image. We could then train the network by providing it with numerous examples of pixels representing parts of walls and floors under various lighting conditions.

### Dictation

1. countless　　　2. Pattern　　　　3. recognition　　　4. classification　　5. vehicles
6. scans　　　　　7. trained　　　　8. written　　　　　9. scenes　　　　　10. self-driving
11. robots　　　　12. heuristically　13. statistical　　　14. stock　　　　　15. judge
16. autonomously　17. algorithms　　18. commercial　　　19. vector　　　　　20. motivated

# Unit 7

## Part 1　Reading & Translating
### Section A

Ⅰ. 1. F　　　　　2. F　　　　　3. F　　　　　4. T　　　　　5. T

Ⅱ. 1. D        2. D           3. C
Ⅲ.
1. comprised   2. depending   3. thinking    4. refers     5. given
6. connected   7. input       8. actually    9. simple     10. likely

IV. LSTM 网络

长短期记忆网络通常被称为 LSTM，是一种特殊的 RNN，它能够学习长期依赖关系。它由 Hochreiter & Schmidhuber(1997)提出，并在随后的工作中被许多人改进和推广。它们在各种各样的问题上表现得非常好，因此现在被广泛使用。

LSTM 是为了避免长期依赖性问题而专门设计的。记住长期信息实际上是这种模型的默认行为，而无须专门学习。

## Section B

Ⅰ. 1. F        2. F           3. F           4. T          5. T
Ⅱ. 1. C        2. D           3. C
Ⅲ.
1. approaches  2. objects     3. Rather      4. only       5. other
6. bigger      7. have        8. generally   9. beyond     10. recurrent

IV. 正则化

正则化是通过惩罚高值回归系数来避免过度拟合的一种方法。简单地说，它减少了参数并缩小(简化)了模型。这种更新型的、更精简的模型可能在预测方面表现更好。正则化可将惩罚加到更复杂的模型中，然后将潜在模型按照过拟合程度从小到大排序，过拟合程度最低的模型通常具有最佳的预测能力。

# Part 3  Listening & Speaking

## Listening Comprehension

1.（B）         2.（C）         3.（A）

**Original**

**Generative Adversarial Network**

A Generative Adversarial Network（GAN）is a type of construct in neural network technology that offers a lot of potential in the world of artificial intelligence. A generative adversarial network is composed of two neural networks: a generative network and a **discriminative** network. These work together to provide high-level simulation of conceptual tasks.

In a generative adversarial network, the generative network constructs results from input, and "shows" them to the discriminative network. The discriminative network is supposed to distinguish between **authentic** and synthetic results given by the generative network.

Experts sometimes describe this as the generative network trying to "fool" the discriminative network, which has to be trained to recognize particular sets of patterns and models. The use of generative adversarial networks is somewhat common in image processing, and in the development of new deep **stubborn** networks that move toward more high-level simulation of human cognitive tasks. Scientists are looking at the potential that generative adversarial networks have to advance the power of neural networks and their ability to "think" in human ways.

## Dictation

1. neural       2. directed      3. exhibit      4. sequence     5. process
6. applicable   7. recognition   8. refer        9. general      10. infinite
11. behavior    12. strictly     13. cyclic      14. stored      15. under
16. replaced    17. incorporates 18. loops       19. gated       20. Short-Term

# Unit 8

## Part 1　Reading & Translating

### Section A

Ⅰ. 1. T　　　2. F　　　3. T　　　4. T　　　5. T
Ⅱ. 1. A　　　2. B　　　3. B
Ⅲ.
1. finds　　　2. involve　　　3. uses　　　4. order　　　5. types
6. learning　　7. strategies　　8. convolutional　9. model　　10. advance

Ⅳ. 马尔可夫决策过程

　　强化学习是机器学习的一种。它允许机器和软件代理自动确定在特定上下文中的理想行为,从而最大限度地提高其性能。简单的奖励反馈是代理学习其行为所必需的,被称为强化信号。

　　有许多不同的算法可以解决这个问题。事实上,强化学习是由一类特定的问题定义的,它的所有解都被归类为强化学习算法。在这类问题中,代理应该根据其当前状态来决定要选择的最佳操作。当这一步骤重复时,这类问题被称为马尔可夫决策过程。

### Section B

Ⅰ. 1. T　　　2. T　　　3. F　　　4. F　　　5. T
Ⅱ. 1. B　　　2. A　　　3. D
Ⅲ.
1. trained　　2. able　　　3. agent　　　4. negative　　5. implicit
6. correct　　7. taken　　　8. base　　　　9. deal　　　10. short

Ⅳ. 什么是"深度"Q-learning?

　　Q-learning 是一个简单但功能强大的算法,可以为我们的代理创建一个备忘录,从而有助于代理确定要执行的操作。

　　但是如果这个备忘录太长呢? 假设一个环境有 10 000 个状态,每个状态有 1000 个操作。这将创建一个包含 1000 万个单元格的表。事态很快就会失控!

　　很明显,我们不能从已经探索的状态中推断出新状态的 Q 值。这带来了两个问题:

　　首先,存储和更新该表所需的内存量将随着状态数的增加而增加。

　　其次,探索每个状态创建所需 Q 表需要的时间是过于庞大的。

　　一个想法应运而生:如果我们用机器学习模型(比如神经网络)来估计这些 Q 值呢? 这正是 DeepMind 算法背后的想法,导致它以 5 亿美元被谷歌收购。

## Part 3　Listening & Speaking

### Listening Comprehension

1.（B）　　　2.（C）　　　3.（C）

**Original**

**Deep Reinforcement Learning**

　　Several of the achievements surrounding Reinforcement Learning（RL）in the past several years are due to the combination of RL with deep learning techniques in **addressing** challenging sequential decision-making problems. This combination, called deep RL, is most useful in problem domains having high dimensional state-space. Deep learning's extension to the domain of RL is considered to be an important technological evolution by many involved in the field.

　　Previous RL approaches led to difficult design issues **with respect to** choice of features. Deep RL, however, has been rather successful in complex tasks with lower prior knowledge thanks to its ability to learn different levels of abstractions from data. For instance, a deep RL agent can successfully learn from visual inputs made up of many thousands of pixels. This presents the potential to mimic some human problem-solving capabilities, even in high dimensional state-space, which was difficult to consider just a few years ago.

Several notable examples of using deep RL in playing games have **stood out** for attaining super-human level expertise. For example, DeepMind (a British AI company owned by Alphabet, Inc.) was able to learn how to play games, reaching human-level performance on many tasks:
- Atari video games
- Go (defeating some of the **toughest** players)
- Poker (beating the world's top professionals)

Deep RL also has the potential for real-world applications such as robotics, autonomous vehicles, healthcare, finance, just to name a few. Nevertheless, several challenges arise in applying deep RL algorithms. For instance, exploring the environment efficiently or being able to generalize what's considered good behavior in a slightly different context are not straightforward to achieve. **In response**, a large collection of algorithms have been proposed for the deep RL framework, depending on a variety of attributes of the sequential decision-making problem domain.

## Dictation

1. popular        2. challenges     3. simulation    4. tends        5. specific
6. preparing      7. relatively     8. autonomous    9. critical     10. loose
11. obstacles     12. problematic   13. Scaling      14. through     15. penalties
16. catastrophic  17. erased        18. optimum      19. optical     20. optimize

# Unit 9

## Part 1  Reading & Translating

### Section A

Ⅰ. 1. F        2. T        3. F        4. T        5. T
Ⅱ. 1. C        2. D        3. D
Ⅲ.
1. by           2. manageable    3. characteristic   4. effectively   5. describing
6. useful       7. needed        8. given            9. facilitate    10. process

Ⅳ. 模式识别

模式识别是利用机器学习算法进行识别模式的过程。模式识别可以定义为基于已经获得的知识或基于从模式及其表示中提取的统计信息的数据分类。模式识别的一个重要方面是它的应用潜力。

在典型的模式识别应用中，原始数据被处理并转换成一种适合机器使用的形式。模式识别涉及模式的分类和聚类。

### Section B

Ⅰ. 1. F        2. F        3. F        4. F        5. T
Ⅱ. 1. B        2. C        3. D
Ⅲ.
1. together     2. approach     3. networks     4. distinguish   5. identify
6. Using        7. like         8. given        9. related       10. training

IV. 深度学习中的图像分割

许多计算机视觉任务需要对图像进行智能分割，以了解图像中的内容，并使对每个部分的分析更容易。现代的图像分割技术使用计算机视觉的深度学习模型，以十年前难以想象的水平，准确地理解图像的每个像素代表了哪个真实世界的对象。

深度学习可以学习视觉输入中的模式，以便预测构成图像的对象类。用于图像处理的主要深度学习模型结构是卷积神经网络（CNN），或特定的 CNN 框架，如 AlexNet、VGG、Inception 和 ResNet。计算机视觉的深度学习模型通常在专门的图形处理单元（GPU）上进行训练和运行，以减少计算时间。

# Part 3    Listening & Speaking

## Listening Comprehension

1．（C）        2．（D）        3．（C）

**Original**

**Pattern Recognition**

Pattern recognition is the automated recognition of patterns and regularities in data. Pattern recognition is closely related to artificial intelligence and machine learning, together with applications such as data mining and Knowledge Discovery in Databases (KDD), and is often used interchangeably with these terms. However, these are distinguished: machine learning is one approach to pattern recognition, while other approaches include **hand-crafted** (not learned) rules or heuristics; and pattern recognition is one approach to artificial intelligence, while other approaches include symbolic artificial intelligence.

Here we focus on machine learning approaches to pattern recognition. Pattern recognition systems are in many cases trained from labeled "training" data (supervised learning), but when no labeled data are available other algorithms can be used to discover previously unknown patterns (unsupervised learning). Machine learning is the common term for supervised learning methods and originates from artificial intelligence, whereas KDD and data mining have a larger focus on unsupervised methods and stronger connection to business use. Pattern recognition has its origins in engineering, and the term is popular in the context of computer vision: a leading computer vision conference is named Conference on Computer Vision and Pattern Recognition. In pattern recognition, there may be a higher interest to formalize, explain and visualize the pattern, while machine learning traditionally focuses on maximizing the recognition rates. Yet, all of these domains have evolved substantially from their roots in artificial intelligence, engineering and statistics, and they've become increasingly similar by integrating developments and ideas from each other.

Pattern recognition algorithms generally aim to provide a reasonable answer for all possible inputs and to perform "most likely" matching of the inputs, **taking into account** their statistical variation. This is **opposed to** pattern matching algorithms, which look for exact matches in the input with pre-existing patterns. A common example of a pattern-matching algorithm is regular expression matching, which looks for patterns of a given sort in **textual data** and is included in the search capabilities of many text editors and word processors. In contrast to pattern recognition, pattern matching is not generally a type of machine learning, although pattern-matching algorithms (especially with fairly general, carefully tailored patterns) can sometimes succeed in providing similar-quality output of the sort provided by pattern-recognition algorithms.

## Dictation

1．subset          2．broad          3．rules          4．trial          5．suitable
6．algorithms     7．predict        8．analogous     9．weights       10．modes
11．adjusted      12．unknown       13．overview     14．common       15．implemented
16．practice      17．fragile       18．tailored     19．pixels       20．lighting

# Unit 10

## Part 1　Reading & Translating

### Section A

Ⅰ. 1. T　　　　2. F　　　　3. T　　　　4. F　　　　5. T
Ⅱ. 1. D　　　　2. B　　　　3. C
Ⅲ.
　1. required　　2. refers　　　3. span　　　4. generally　　5. retrieving
　6. particular　7. randomly　　8. used　　　9. recognition　10. creation
Ⅳ. 机器翻译

　　机器翻译系统是一类应用程序或在线服务，使用机器学习技术将大量文本从其支持的语言中翻译出来或翻译成它们支持的语言。该服务将"源"文本从一种语言翻译成另一种"目标"语言。

　　尽管机器翻译技术背后的概念和使用它的接口相对简单，但背后的科学和技术却极其复杂，它汇集了一些前沿技术，特别是深度学习（人工智能）、大数据、语言学、云计算和 Web API。

### Section B

Ⅰ. 1. F　　　　2. T　　　　3. F　　　　4. T　　　　5. T
Ⅱ. 1. D　　　　2. B　　　　3. D
Ⅲ.
　1. source　　　2. designed　　3. using　　　4. stands　　　5. connected
　6. based　　　7. sequentially　8. enabled　　9. known　　　10. related
Ⅳ. Word2vec

　　Word2vec 是一个处理文本的双层神经网络。它的输入是一个文本语料库，输出是一组向量，即该语料库中单词的特征向量。虽然 Word2vec 不是一种深度神经网络，但它可将文本转换为深度网络可以理解的数值形式。Deeplearning4j 是 Java 和 Scala 实现了 Word2vec 的一种分布式形式，它在 Spark 上通过 GPU 运行。

　　Word2vec 的应用不仅仅是解析句子。它也可以应用于基因、代码、喜好、播放列表、社交媒体图和其他可以识别模式的语言或符号序列。

## Part 3　Listening & Speaking

### Listening Comprehension

　1.（D）　　　　2.（B）　　　　3.（C）

**Original**

**Speech Recognition**

　　Speech recognition is the inter-disciplinary sub-field of computational linguistics that develops methodologies and technologies that enables the recognition and translation of spoken language into text by computers. It is also known as Automatic Speech Recognition (ASR), computer speech recognition or Speech To Text (STT). It incorporates knowledge and research in the linguistics, computer science, and electrical engineering fields.

　　Some speech recognition systems require "training" (also called "**enrollment**") where an individual speaker reads text or isolated vocabulary into the system. The system analyzes the person's specific voice and uses it to fine-tune the recognition of that person's speech, resulting in increased accuracy. Systems that do not use training are called "speaker independent" systems. Systems that use training are called "speaker dependent".

　　Speech recognition applications include voice user interfaces such as voice dialing (e.g. "call home

"), **call routing** (e.g. "I would like to make a **collect call**"), **demotic** appliance control, search (e.g. find a podcast where particular words were spoken), simple data entry (e.g., entering a credit card number), preparation of structured documents (e.g. a **radiology** report), determining speaker characteristics, speech-to-text processing (e.g., word processors or emails), and aircraft (usually **termed** direct voice input).

The term voice recognition or speaker identification refers to identifying the speaker, rather than what they are saying. Recognizing the speaker can simplify the task of translating speech in systems that have been trained on a specific person's voice or it can be used to **authenticate** or verify the identity of a speaker as part of a security process.

From the technology perspective, speech recognition has a long history with several waves of major innovations. Most recently, the field has benefited from advances in deep learning and big data. The advances are **evidenced** not only by the surge of academic papers published in the field, but more importantly by the worldwide industry adoption of a variety of deep learning methods in designing and deploying speech recognition systems.

## Dictation

1. Interpretation
2. subtopic
3. deals
4. comprehension
5. considered
6. considerable
7. automated
8. categorization
9. archiving
10. algorithms
11. utilizes
12. misrecognized
13. profiles
14. discern
15. fragmented
16. probability
17. known
18. derives
19. shown
20. consumed

# Bibliography

[1] Dale N, Lewis J. Computer Science Illustrated[M]. 7th ed. Burlington: Jones & Bartlett Learning, 2019.
[2] Forouzan B. Foundations of Computer Science[M]. 4th ed. Boston: Cengage Learning EMEA, 2018.
[3] Russell S J, Norvig P. Artificial Intelligence A Modern Approach[M]. 3rd ed. New York: Pearson Education, Inc., 2010.
[4] Ertel W. Introduction to Artificial Intelligence[M]. 2nd ed. New York: Springer International Publishing AG, 2017.
[5] 吕云翔. 计算机专业英语[M]. 北京: 电子工业出版社, 2018.

# 图书资源支持

感谢您一直以来对清华版图书的支持和爱护。为了配合本书的使用,本书提供配套的资源,有需求的读者请扫描下方的"书圈"微信公众号二维码,在图书专区下载,也可以拨打电话或发送电子邮件咨询。

如果您在使用本书的过程中遇到了什么问题,或者有相关图书出版计划,也请您发邮件告诉我们,以便我们更好地为您服务。

**我们的联系方式:**

地　　址: 北京市海淀区双清路学研大厦 A 座 714

邮　　编: 100084

电　　话: 010-83470236　010-83470237

客服邮箱: 2301891038@qq.com

QQ: 2301891038（请写明您的单位和姓名）

**资源下载:** 关注公众号"书圈"下载配套资源。

资源下载、样书申请

书　圈

获取最新书目

观看课程直播